Troubleshooting the Sequencing Batch Reactor

WASTEWATER MICROBIOLOGY SERIES

Editor

Michael H. Gerardi

Nitrification and Denitrification in the Activated Sludge Process
Michael H. Gerardi

Settleability Problems and Loss of Solids in the Activated Sludge Process
Michael H. Gerardi

The Microbiology of Anaerobic Digesters
Michael H. Gerardi

Wastewater Pathogens
Michael H. Gerardi and Mel C. Zimmerman

Wastewater Bacteria
Michael H. Gerardi

Microscopic Examination of the Activated Sludge Process
Michael H. Gerardi

Troubleshooting the Sequencing Batch Reactor
Michael H. Gerardi

Troubleshooting the Sequencing Batch Reactor

Michael H. Gerardi

Illustrations by Brittany Lytle

A John Wiley & Sons, Inc., Publication

Library of Congress Cataloging-in-Publication Data:

Gerardi, Michael H.
    Troubleshooting the sequencing batch reactor/Michael H Gerardi; illustrations by Brittany Lytle.
      p. cm.
    Includes index.
    ISBN 978-0-470-05073-6 (pbk.)
  1. Sewage–Purification–Sequencing batch reactor process. I. Title.
    TD756.G47 2010
    628.3′2—dc22

                              2010006498

10 9 8 7 6 5 4 3 2 1

To
Allan N. Young, Jr.
and
the men and women of Cromaglass® Corporation

*The author extends his sincere appreciation to
Brittany Lytle for artwork used in this text.*

Contents

Preface

The sequencing batch reactor (SBR) is a modification of the activated sludge process that uses a "fill-and-draw" mode of operation. There are several operational and economic advantages for using SBR technology in lieu of conventional, continuous-flow mode of operation for the treatment of wastewater. However, the successful operation of SBR involves more in-depth knowledge of the activity of the biomass than does the continuous-flow mode of operation. This knowledge enables the operator to (1) modify the operation of the SBR for optimal biomass activity and (2) troubleshoot the SBR to identify problematic conditions and establish proper, process control measures for cost effective operation, and permit compliance.

Troubleshooting of any biological wastewater treatment process involves an in-depth review, correlation, and evaluation of much data, including the flow, mode of operation, industrial discharges, supportive sampling with acceptable collection points and sampling procedures, timely and accurate laboratory analyses, calculated operational parameters, and knowledge of bacteria and bacterial activity. This book provides an in-depth review of the bacteria and bacterial activity involved with SBR technology.

In-depth knowledge of the biomass may be obtained from the operation and maintenance (O & M) manual, on-site manufacture's training, public and private short courses, and appropriate literature. This book provides (1) basic and in-depth reviews of the bacteria and their activities in SBR that occur during aerobic, anoxic, and anaerobic/fermentative conditions, (2) the operational tools—biological, chemical, and physical—that are needed to monitor acceptable and unacceptable activity, and (3) the control measures needed for cost-effective operation and permit compliance. Because an operator has limited ability to control the wastewater strength and composition, an operator's ability to monitor and regulate bacterial activity is critical to the success of the SBR.

This book contains numerous illustrations of acceptable and unacceptable operational conditions, troubleshooting keys and tables for the identification of unacceptable conditions, and recommendations for correcting unacceptable conditions. *Troubleshooting the Sequencing Batch Reactor* is the seventh book in the Wastewater Microbiology Series by John Wiley & Sons. The series is designed for wastewater personnel, and the series presents a microbiological review of the significant groups of organisms and their roles in wastewater treatment facilities.

Linden, Pennsylvania   MICHAEL H. GERARDI

Part I

Overview

1

Introduction

The sequencing batch reactor (SBR; see Figure 1.1) is a suspended-growth, wastewater treatment process. It is a modification of the activated sludge process (Figure 1.2) and may be described as simply a holding tank for receiving a batch of wastewater for treatment. Once the batch is treated, a portion of the batch is discharged and another batch of wastewater is collected, treated, and discharged and another batch may then be sequentially collected, treated, and discharged. There are two classifications of SBR: the intermittent flow or "true batch reactor" (Figure 1.3) and the continuous flow (Figure 1.4). The intermittent flow SBR may be operated as a single-feed or a multiple-feed reactor (Figure 1.5).

The intermittent-flow SBR accepts wastewater or influent only at specified intervals and uses time sequences or five phases over a cycle (Figure 1.6) to perform numerous treatment operations that the conventional, activated sludge process performs in numerous tanks. There are usually two reactors in parallel. Because one reactor is closed to influent during the treatment of a batch of wastewater, two reactors may be operated in parallel with one reactor receiving influent while the other reactor operates through its cycle of phases. The cycle of the SBR can be designed or modified to (1) vary operational strategy to provide for aerobic, anoxic, and anaerobic/fermentative conditions and proliferation of desirable bacteria and (2) enhance the removal of ammonia, nitrogen, and phosphorus.

Modifications or changes in phases permit the SBR to treat fluctuating quantities and compositions of wastewater while maintaining a high-quality effluent or decant. The intermittent-flow SBR may be filled once with wastewater to its normal operating level, and the wastewater then is treated through all phases of a cycle. The intermittent-flow SBR may also be filled several small batches of wastewater until the normal operating level is reached. However, after each small batch is placed in the SBR, the batch is treated or aerated before the next small batch is placed in the

Troubleshooting the Sequencing Batch Reactor, by Michael H. Gerardi
Copyright © 2010 by John Wiley & Sons, Inc.

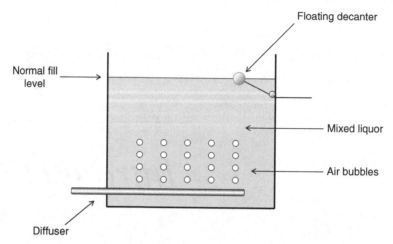

Figure 1.1 Sequencing batch reactor. The typical sequencing batch reactor consists of a rectangular or square basin. Aeration and mixing is provided with fine air diffusers. After completion of a Fill Phase, a React Phase, and a Settle Phase, the supernatant or decant is removed during the Decant Phase with a floating decanter. In the sequencing batch reactor, one basin serves as the aeration tank for a period of time and then serves as the sedimentation basin or clarifier for a period of time.

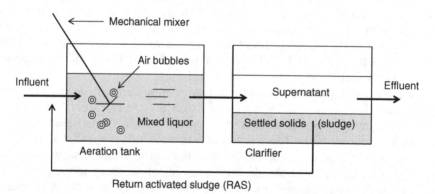

Figure 1.2 Conventional, activated sludge process. The conventional, activated sludge process consists of at least one upstream tank, the aeration tank, and at least one downstream tank, the sedimentation basin or clarifier. Although the clarifier provides for the separation and settling of solids from the suspending medium, it differs greatly from the sequencing batch reactor, because a continuous flow of wastewater enters the clarifier and a return pump is required to remove the settled solids from the clarifier and return them (return activated sludge or RAS) to the aeration tank. In the aeration tank, aeration may be provided by coarse or fine air mechanism systems, and mixing may be provided through aeration or with a mechanical mixer.

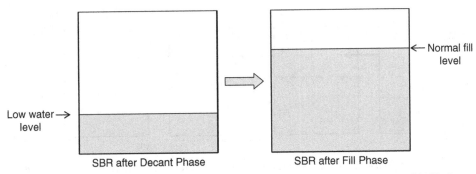

Low water → level

← Normal fill level

SBR after Decant Phase

SBR after Fill Phase

Figure 1.3 *SBR, intermittent flow. Typically, the intermittent flow or "true batch reactor" is filled once with wastewater to its normal fill level, and the wastewater is then treated. After filling the reactor to its normal fill level, no additional wastewater is added to the sequencing batch reactor until all phases have been completed and sufficient decant has been removed to permit the discharge of another batch of wastewater to the reactor.*

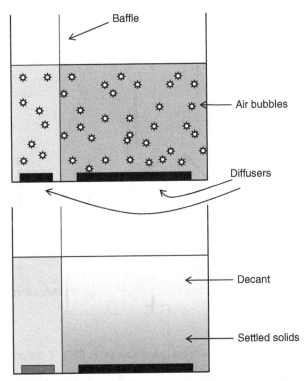

Baffle

Air bubbles

Diffusers

Decant

Settled solids

Figure 1.4 *SBR, continuous flow. In the continuous-flow, sequencing batch reactor, influent always enters the reactor. There are two chambers in the reactor that are divided by a baffle. The smaller chamber receives the influent, and from here the influent slowly moves into the larger chamber. The larger chamber acts as the sequencing batch reactor. However, the sequencing batch reactor only has a limited number of phases: React, Settle, and Decant.*

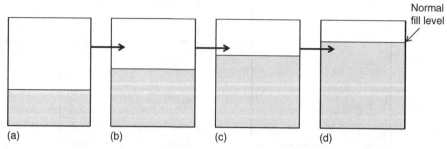

Figure 1.5 SBR, multiple flow. In the multiple-feed, sequencing batch reactor, there are several fill periods before the normal fill level is reached. After the Decant Phase the sequencing batch reactor is at its normal low water level (**a**) and the first batch of wastewater is discharged to the sequencing batch reactor (**b**). After this batch of wastewater is received, the sequencing batch reactor then enters a React Phase. After the React Phase a second batch of wastewater is discharged to the sequencing batch reactor (**c**) and the React Phase is repeated. Again, an additional batch of wastewater is discharged to the sequencing batch reactor (**d**), and the React Phase is repeated once more. This process of multiple feeds is continued until the sequencing batch reactor is at its normal fill level. From this level the sequencing batch reactor would enter another React Phase, then Settle Phase, and finally Decant Phase.

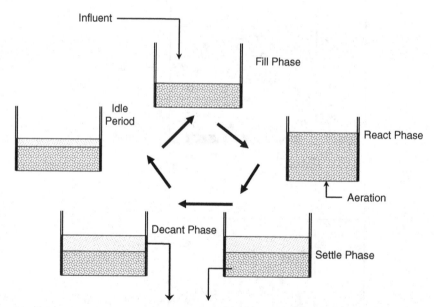

Figure 1.6 Five phases of an SBR. There are five phases of an SBR. These phases consist of the Fill Phase, the React Phase, the Settle Phase, the Decant Phase, and the Idle Phase or Idle Period. During the Fill Phase, influent is discharged to the SBR. The Fill Phase may consist of an Aerated Fill, Mix Fill, and/or Static Fill. Once the normal fill level is reached, the SBR enters the React Phase or aerated period of the cycle. After aeration, the SBR enters the Settle Phase, where a quiescent condition is established (no aeration and no mixing) and solids settle in the reactor to produce a high-quality supernatant or decant. After the Settle Phase, supernatant or decant is removed during the Decant Phase. Wasting of solids may be performed during the React Phase, the Settle Phase, or the Decant Phase. If time permits before the start of the next Fill Phase, the SBR may be "parked" or placed in an Idle Phase or, more appropriately, Idle Period.

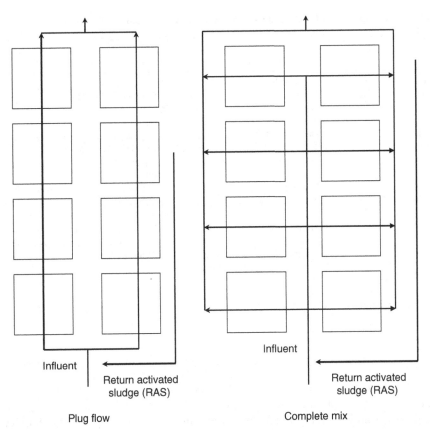

Figure 1.7 *Modes of operation of the conventional, activated sludge process. Although there are several modes of operation of the conventional, activated sludge process, two of the more commonly used modes are (1) plug flow and (2) complete mix. In the plug flow mode of operation, influent wastewater is mixed with return activated sludge (RAS) and then discharge in "train-like" fashion through a series of aeration tanks. Plug flow mode of operation established a nutrient gradient and substrate (food) gradient throughout all tanks that are in-line. In the complete mix mode of operation, influent wastewater is mixed with return activated sludge (RAS) and then discharged equally through all aeration tanks in-line. In the complete mix mode of operation, substrate and toxic components are diluted in each aeration tank.*

SBR. This mode of operation is referred to as multiple or step feed and is used to treat wastewater having high-strength BOD.

In the continuous-flow SBR, influent flows continuously during all phases of the cycle. To reduce short-circuiting of pollutants, a partition or baffle typically is added to the SBR to separate the turbulent aeration zone from the quiescent zone.

SBR operate on a fill-and-draw (batch feed and batch draw or decant) method for the treatment of wastewater. The fill-and-draw method was developed in the early 1900s in the United Kingdom, has been widely used in the United States, Canada, and Europe since the 1920s and has undergone much improvement since the 1950s. However, due to the high degree of operator attention and automation required by SBR as well as the clogging of aeration diffusers when aeration blowers shut off for the periodic settling of solids, the "conventional," activated sludge process was favored over the SBR. Although the conventional, activated sludge process can be operated in several different modes (Figure 1.7), the SBR

TABLE 1.1 Comparison of SBR and Conventional, Activated Sludge Process

	Treatment Process	
Parameter	Sequencing Batch Reactor	Conventional, Activated Sludge Process
Influent	Periodic feed	Continuous feed
Discharge	Periodic discharge	Continuous discharge
Organic loading	Cyclic	Continuous
Aeration	Intermittent	Continuous
Mixed liquor	Reactor only	Aeration tank and clarifier; Recycled from tank to clarifier
Clarification	Ideal, no influent flow	Nonideal, influent flow from aeration tank
Flow pattern	Plug-flow	Complete mix, approaching plug-flow or other
Equalization of flow	Yes	No
Flexibility	Adjustment of aerobic, anoxic, and anaerobic/ fermentative periods as well as settling period	Limited ability to adjust aerobic, anoxic, and anaerobic/fermentative periods of settling period
Clarifier required	No	Yes
Return sludge required	No	Yes

combines all treatment steps into a single tank whereas the conventional, activated sludge process relies on multiple tanks (Table 1.1). In the 1970s a pre-react selection period (anoxic and/or anaerobic/fermentative time period during the Fill phase) was incorporated in the SBR to control undesired filamentous organism growth. The selection period along with modern aeration equipment and computer control systems has advanced the use of the SBR.

The key to the SBR process is the control system. The system contains a combination of level sensors, timers, and microprocessors that provide flexibility and accuracy in operating the SBR. By varying the phase times for aerobic (oxic), anoxic, and anaerobic/fermentative of a given sequence or cycle, the biological reactions for nitrification, denitrification, and biological phosphorus removal can be controlled.

SBR are used to treat domestic, municipal, and industrial wastewaters, particularly in areas that have low flows or highly variable flow patterns. The use of SBR technology has grown rapidly in small communities that produce less than 1 million gallons per day (MGD). On-site, sequencing batch reactors are ideally suited for use in single-family homes, farms, hotels, small businesses, casinos, and resorts, where centralized wastewater treatment facilities do not exist. Most recently, more and more small communities with up to 10 MGD are using SBR technology to reduce capital expenses and operation and maintenance (O & M) costs and to comply with more stringent effluent requirements, including nutrient removal (Table 1.2). However, there are several disadvantages of SBR, including significant head loss through the system, difficulty in removing floating materials, and intermittent decant that generally requires equalization before downstream treatment processes such as filtration and disinfection (Table 1.3).

TABLE 1.2   **Advantages of the SBR as Compared to the Conventional Activated Sludge Process**

Anoxic period (mix fill phase) provides for alkalinity recovery
Anoxic period (mix fill phase) provides for better settling floc particles due to the control of undesired filamentous organism growth
Complete quiescent, automatic operation for improved total suspended solids (TSS) removal
Elimination of secondary clarifiers and sludge return pumps
Flexible, adaptable, automatic operation
High degree of automation reduces operational staff requirements
Higher mixed liquor temperatures provide for improved bacterial kinetics
Inherent nitrogen removal capability
Inherent phosphorus removal capability
Internal flow equalization
Less process equipment to maintain
Low land requirement and little yard plumbing, since there is no secondary clarifier
Operation flexible to easily change mode of operation
Reduction in sensitivity to constituent concentration surges, that is, no flow surges

TABLE 1.3   **Disadvantages of the SBR as Compared to the Conventional Activated Sludge Process**

Frequent stop/start process machinery
Higher level of control sophistication (knowledgeable operators), especially for adjustments in cycle and phase times
Higher maintenance cost due to automated controls
Requires more head drop through plant due to changing liquid level
Two or more basins or a pre-equalization tank for process operation and redundancy

OPERATIONAL COSTS

Major costs associated with the operation of the SBR are (1) electrical consumption (aeration), (2) sludge handling and disposal, and (3) chemicals. Although the bulk of electrical consumption is for aeration (cBOD degradation and nitrification), electrical consumption also is required for the operation of (1) headworks, (2) primary clarifiers, (3) thickener, (4) effluent filters, (5) disinfection, (6) heating, (7) lighting, and (8) post aeration.

Aeration of the SBR is the largest electrical expenditure and is influenced by (1) mean cell residence time (MCRT), especially high MCRT and endogenous respiration; (2) degradation of cBOD—1.8 pounds O_2 per pound cBOD degraded; (3) nitrification—4.6 pounds O_2 per pound ammonium (NH_4^+) oxidized completely to nitrate (NO_3^-); (4) time of aeration; and (5) dissolved oxygen requirement.

SLUDGE HANDLING AND DISPOSAL

Sludge handling and disposal costs are influenced by (1) MCRT, (2) polysaccharide production through nutrient deficiency and Zoogloeal growth or viscous floc, (3) type of thickening and dewatering equipment used, and (4) sludge disposal options (landfill, hazardous waste landfill, incineration, agricultural utilization, and composting).

CHEMICALS

Operational costs associated with chemical applications are influenced by (1) primary treatment requirements, (2) polymer addition to the SBR, (3) coagulant (metal salt) addition to the SBR, (4) nutrient addition, (5) malodor control, (6) pH control, (7) foam control, (8) bioaugmentation, (9) disinfection, (10) phosphorus precipitation, and (11) alkalinity addition.

GENERAL OPERATIONAL AND SYSTEM SIZING GUIDELINES

To obtain ideal operational conditions for wastewater treatment facilities using SBR technology for intermittent feed process, the following guidelines for general operation and system sizing are offered:

- At least three reactors should be available.
- Cycle times should be based on the design maximum daily flow.
- Facilities should be available for the equalization of flows and organic slug discharges.
- Design food-to-microorganism (F/M) ratios and mixed liquor suspended solids (MLSS) concentrations should be similar to other conventional and extended aeration processes. MLSS should be 2000 to 3000 mg/L. For the treatment of domestic wastewater with a nitrification requirement, the F/M should be 0.05 to 0.1. For the treatment of domestic wastewater without a nitrification requirement, the F/M should be 0.15 to 0.4.
- Reactor MLSS and mixed liquor volatile suspended solids (MLVSS) concentrations should be calculated at the low-water level.
- The low-water level should be >10 ft.
- Treatment tanks downstream of the SBR should be sized to handle the peak discharge rate.
- Sampling procedures for each SBR should consider process control as well as compliance reporting.
- For biological phosphorus release the SBR should have dissolved oxygen <0.8 m/L and nitrates (NO_3^-) <8 mg/L, and substrates should be available as soluble cBOD, especially fatty acids.
- For denitrification the SBR should have dissolved oxygen <0.8 m/L.

2

SBR Cycles

The number of cycles and times of phases in each cycle of a sequencing batch reactor (SBR) are determined by (1) the quantity of wastewater or flow to be treated, (2) the strength of the wastewater, (3) the treatment requirements, and (4) the number of parallel reactors available for use at the wastewater treatment plant (Figure 2.1). Increasing flow results in decreased treatment time and an increase in the number of cycles per day with decreasing time per cycle. Increasing strength of wastewater requires more treatment time and a decrease in the number of cycles per day with increasing time per cycle, provided that an adequate number of reactors are available to satisfy the increase in treatment time. An effluent discharge requirement for nitrogen and/or phosphorus often requires an increase in treatment time and a decrease in the number of cycles per day.

An increase in the number of reactors at a wastewater treatment plant permits more treatment time per cycle or a smaller number of cycles per day for each reactor. Typically, four cycles (six hours per cycle) to six cycles (four hours per cycle) per day are used to treat domestic and municipal wastewater. The operational factors that determine the number of cycles per day are the volume of wastewater to be treated and the minimum treatment time required. Fill Phase times must be equal to non-Fill Phase times for dual reactor systems.

Some sequencing batch reactors may be operated with an odd number of cycles per day—for example, 4.5 cycles. This is done in order to change slightly the daily loading and loading sequence applied to each reactor according to variations in flow (Figure 2.2).

The time allotted to each phase of a cycle can vary greatly (Figure 2.3). The Fill Phase may be 25% to 50% of the cycle time, but the higher percentage of time for the Fill Phase may include (1) Static Fill or an anaerobic/fermentation time, (2) Mix Fill or an anoxic time, and (3) React Fill or aerobic time. React Phase may be 15%

Troubleshooting the Sequencing Batch Reactor, by Michael H. Gerardi
Copyright © 2010 by John Wiley & Sons, Inc.

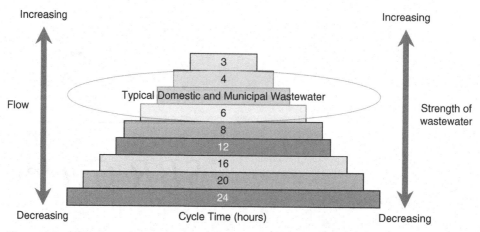

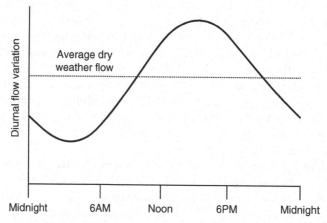

Figure 2.1 *SBR cycles. The number of cycles per day that are operated in an SBR is dependent upon (1) wastewater flow and (2) wastewater strength. With increasing flow, the number of cycles per day decreases and with decreasing strength the number of cycles per day decreases. Typically, four to six cycles per day are used for the treatment of domestic and municipal wastewater.*

Figure 2.2 *Variations in flow. Wastewater flow to an SBR varies greatly through the day for domestic and municipal wastewater treatment plants. This is due to the diurnal flow pattern of residential, commercial, and industrial discharges to the sewer system. Peak hydraulic loading usually occurs between 10 AM and 6 PM, while lowest hydraulic loading usually occurs between 10 PM and 6 AM.*

to 20% of the cycle time, and Settle Phase may be 15% to 25% of the cycle time. Draw Phase or Decant Phase can range from 5% to 50% of the cycle time. However, the greater the percentage of the cycle time that is Settle Phase and Decant Phase, the greater the need to troubleshoot operational conditions responsible for poor settling of solids or loss of solids from the sludge blanket.

The mode of operation of an SBR typically is time-paced, but the operation may be flow-paced. During a flow-paced mode of operation, each SBR receives the same

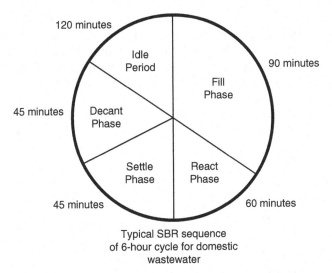

Typical SBR sequence
of 6-hour cycle for domestic
wastewater

Figure 2.3 *Phase times per cycle. Although phase times per cycle vary according to the number of reactors in-line, wastewater flow, and wastewater strength, typical periods of time for each phase in a 6-hr cycle consist of approximately 30 min for Fill Phase, 60 min for React Phase, 45 min for Settle Phase, and 45 min for Decant Phase. This would leave approximately 120 min for an Idle Phase or Idle Period.*

quantity of flow or hydraulic loading and approximately the same quality or composition and strength of organic loading during each cycle. During a time-paced mode of operation, each SBR receives a different hydraulic loading and a different strength of organic loading during each cycle. Unless the time cycle for each SBR is adjusted, the changes in hydraulic and organic loading conditions during the cycle make it more difficult than flow-paced mode of operation to operate, especially if denitrification is required. Time-paced mode of operation may not provide adequate soluble cBOD or carbon for denitrification.

There are two environmental factors that affect the number of cycles per day and the length of phases per cycle. These factors are cold weather and wet weather.

COLD WEATHER OPERATION

Cold weather—or, more appropriately, cold wastewater temperature—severely impacts wastewater treatment efficiency. Changes in wastewater temperature influence the activity of the bacteria in an SBR and consequently influence the rate of specific treatment objectives such as floc formation and nitrification. Cold temperatures also promote the growth of the foam-producing filamentous organism *Microthrix parvicella*.

As temperature increases, biological or enzymatic activity increases, until enzymes are damaged (denatured) and treatment efficiency is lost. As temperature decreases, enzymatic activity decreases until enzymes are inhibited and treatment efficiency again is lost. For every 10 °C increase in temperature, the rate of enzymatic activity doubles.

The optimum temperature for the operation of an SBR is 20 °C to 25 °C. Therefore, when decreasing temperatures or depressed temperatures (<16 °C) occur, appropriate operational measures should be taken to maintain effective wastewater treatment and permit compliance. This is of great importance to wastewater treatment facilities with small flows, especially those that receive very little flow or perhaps no flow between 11:00 PM and 6:00 AM.

Operational measures that may be used to maintain efficient treatment during depressed wastewater temperatures include:

- Increase MLVSS
- Add bioaugmentation products
- Add coagulant and/or polymer
- Increase reaction time for specific phases
- Place additional reactors in-line
- Reduce influent cBOD loading to the SBR
- Increase dissolved oxygen concentration

A combination of operational measures may be needed to achieve efficient treatment, if one measure alone is not sufficient.

Increase MLVSS

If bulking does not occur during the Settle Phase or Decant Phase, then the concentration of MLVSS may be increased. This would provide for more cBOD-removing (organotrophic) bacteria and more nitrifying bacteria. Although each group of bacteria become sluggish during depressed wastewater temperatures, the presence of a larger population of each group of bacteria would permit more rapid degradation of cBOD and more time to nitrify with a larger population of nitrifying bacteria. By increasing the concentration of MLVSS, the F/M ratio decreases.

Add Bioaugmentation Products

Bioaugmentation products consist of commercially prepared cultures of cBOD-removing bacteria, nitrifying bacteria, and some fungi. The cultures are commonly added to biological wastewater treatment systems to achieve specific goals, including:

- Control foam and scum production
- Control undesired growth of filamentous organisms
- Control malodor production
- Improve methane production
- Improve cold weather operation
- Overcome inhibition
- Recovery from toxicity
- Reduce sludge production

If conventional operational measures such as increasing MLVSS concentration to improve cold weather treatment efficiency are not possible or not adequate, then bioaugmentation products may be added to the SBR to provide for more cBOD-removing bacteria and more nitrifying bacteria. The addition of the cultures would mimic an increase in MLVSS by increasing the number of desirable bacteria without increasing solids inventory or MLVSS and the risk of sludge bulking.

Add Coagulant and/or Polymer

A coagulant or metal salt such as lime ($Ca(OH)_2$) or ferric chloride ($FeCl_3$) may be added to the SBR during the React Phase to capture and thicken solids and improve settleability. The coagulant would provide for dense and heavy floc particles that would possess desirable settling qualities. The captured solids would keep more bacteria in the SBR to improve treatment efficiency, and a decrease settling time would allow for more treatment during the React Phase or other phases.

A polymer such as a cationic polyacrylamide polymer may be added to the SBR during the React Phase to capture and thicken solids. The polymer would provide for dense and heavy floc particles. The captured solids would keep more bacteria in the SBR and provide for a decreased settling time. Properly matched or jar-tested coagulants and polymers may be added to the SBR during the React Phase to provide for increased capture and thickening of solids.

Increase Reaction Time for Specific Phases

The reaction time of each phase should be carefully reviewed to determine if the time of any phase should be increased. For example, should the time of the React Phase be increased to improve cBOD removal or improve nitrification? Should the time of the Mixed Fill Phase be increased or the time of the anoxic period after React Phase and Settle Phase increased to provide for improved denitrification? An increase in reaction time of any phase requires a reduction in time of another phase, decrease in number of cycles per day, or the use of another reactor.

Place Additional Reactors In-Line

If additional reactors are off-line and available for use, they should be placed in-line. This would provide for more cycles per day and longer phase times in each cycle for improved treatment efficiency during depressed wastewater temperatures.

Reduce Influent cBOD Loading to the SBR

By reducing the cBOD loading to the SBR, less dissolved oxygen demand and less treatment time are required. Therefore, achieving efficient cBOD reduction and nitrification would be more easily achieved.

Reduction in cBOD loading to the SBR can be achieved through several operational measures, including:

- Reduce significant discharges of fats, oils, and grease (FOG). As much as possible, the FOG should be removed physically at appropriate establishments and disposed at an appropriate landfill or by an alternate means.

• If available, place addition primary clarifiers in-line. Additional clarification capacity or the use of a coagulant and/or polymer with in-line clarifiers provides for reduction in colloidal and particulate cBOD loading to the SBR.

Increase Dissolved Oxygen Concentration

Increasing dissolved oxygen concentration during the React Phase may help to promote more rapid removal of cBOD and improve nitrification. If the cBOD-removing bacteria can remove cBOD more quickly with increased dissolved oxygen concentration, this would provide for more time during the React Phase for nitrification. Unless mechanical conditions or increased loading conditions prevent an increase in dissolved oxygen concentration, an increase in dissolved oxygen concentration is feasible during depressed wastewater temperature due to the greater affinity of cold wastewater for dissolved oxygen as compared to warm wastewater. A dissolved oxygen concentration of at least 3 mg/L is recommended during depressed temperatures.

Consideration should be given to several design features for SBR that are located in cold temperate regions. Reactors for small treatment facilities should be located in housings or coverings to prevent wind chill and freezing. Larger reactors should be covered to prevent wind chill and freezing or insulated with earthen-banks, mulch, or saw dust. Exposed piping should be wrapped with heat tap and insulated, and similar consideration should be given for decanter valves and chemical feed lines.

WET WEATHER OPERATION

Wastewater treatment facilities that have combined sanitary and storm sewers and no combined sewer overflow (CSO) measures are subject to significant inflow. These facilities may also experience significant leakage of water from off-set joints in sewers and laterals, broken sewers and laterals, poorly constructed and maintained manholes, and open pick holes in manholes (infiltration). Until inflow and infiltration (I/I) are corrected, SBR may experience hydraulic overloads that reduce treatment efficiency and may cause permit violations.

I/I problems have several significant impacts upon an SBR, including (1) increased dissolved oxygen concentration, (2) diluted BOD, and (3) cold temperature. The increased dissolved oxygen concentration is of special concern to operators of SBR that are required to remove nitrogen and phosphorus. The increased dissolved oxygen concentration affects the ability of the SBR to produce anoxic and anaerobic conditions in an SBR during Mixed Fill Phase and Static Fill Phase, respectively.

3

SBR Phases

Although the conventional, continuous-flow activated sludge process was developed in part to eliminate the problem of clogging aeration diffusers in the sequencing batch reactor (SBR) by separating substrate removal during aeration from solids separation during clarification (Figure 3.1), many wastewater treatment applications are not appropriate for the continuous-flow process or are better served by batch treatment (Table 3.1). Some continuous-flow processes are too costly due to climatic or regulatory requirements. Some applications served by batch treatment provide significant savings with capital investment and energy, operations, and maintenance costs. Examples include (1) batch treatment of industrial wastewater with priority pollutants, high organic loadings, or transient flows, (2) batch treatment of wastewater to achieve nitrogen and/or phosphorus removal requirements, (3) batch treatment of small (<10 MGD) domestic wastewater flow, and (4) batch treatment of wastewater in temperate regions of the world where cold weather clarifier operations are problematic. Perhaps the major advantage of the batch process compared to the continuous-flow process is its flexibility to modify reactor conditions through time controls and/or dissolved oxygen settings. Due to its operational flexibility, it is rather easy to increase treatment efficiency of an SBR by changing the duration of phases rather than adding or removing treatment tanks as is necessary in the continuous-flow process.

However, in order to successfully use SBR technology two requirements must be satisfied. These requirements are (1) a higher level of control sophistication and operator training and (2) two or more reactors or pre-equalization tanks for process operation and redundancy.

The treatment principles of the SBR are very similar to those of the conventional, activated sludge process, except a secondary clarifier is not required in the SBR. In the conventional process, treatment is performed simultaneously in separate tanks

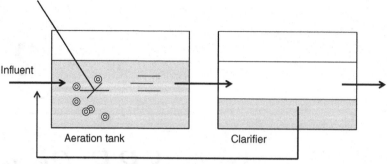

Return activated sludge (RAS)

Figure 3.1   *Conventional, continuous-flow activated sludge process. In the conventional, continuous-flow activated sludge process, aeration of wastewater is separated from settling of solids or clarification of suspended solids. In the upstream tank or aeration tank, bacteria and wastes (solids) are constantly suspended and do not plug aeration equipment. The suspended solids from the aeration tank are discharged to a downstream tank or clarifier where solids are separated from the suspending medium. Here, there is no aeration equipment that can be plugged by the settled solids.*

TABLE 3.1   Benefits of SBR Technology

Complete quiescent settling for improved total suspended solids (TSS) removal.
No separate clarifier.
Capacity upgrades and phasing do not require modification or interruption of current treatment process.
Automation reduces operational staff requirements.
Significantly smaller footprint requires less site work and yard plumbing.
Lower initial capital costs.
Older wastewater treatment plants can be retrofitted to SBR technology because basins are already present.

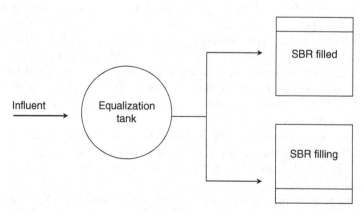

Figure 3.2   *Wastewater treatment plant using two SBR. Typically, wastewater treatment plants using SBR technology have at least two SBR. Influent (raw wastewater) may enter an equalization tank where the influent is then discharged to one SBR under the Fill Phase, while no influent is discharge to the SBR that is filled and undergoing completion of its cycle (React Phase, Settle Phase, and Decant Phase).*

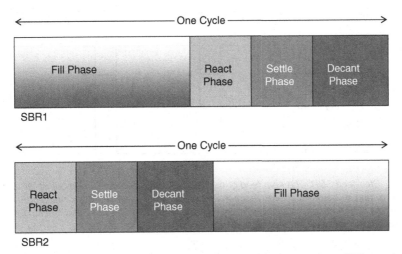

Figure 3.3 *Timing cycles and phases of two SBR. The timing of cycles of two SBR must be established to ensure that as one SBR is in its React Phase, Settle Phase, and Decant Phase (SBR 1) and receiving no influent, the other SBR (SBR 2) is in its Fill Phase. The Fill Phase may consist of Aerated Fill, Mix Fill, and/or Static Fill.*

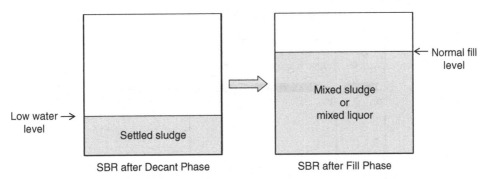

Figure 3.4 *Single-feed SBR. In the single feed (intermittent feed or "true SBR") the SBR is filled once during the Fill Phase from its low water level to its normal fill level. The Fill Phase may consist of Aerated Fill, Mix Fill, and/or Static Fill. The single-feed SBR is used for nitrification.*

(aeration and clarifier), whereas in the SBR, treatment takes place sequentially in a single or common tank; that is, a period of time in the tank is allotted for aeration and mixing and a period of time is allotted for settling of solids without aeration or mixing.

Wastewater treatment plants using SBR technology typically have two reactors that operate in parallel (Figure 3.2). Each SBR operates with a fixed number of cycles per day and the same phases for each cycle (Figure 3.3). The number of cycles each day may be adjusted, and the time of each phase also may be adjusted to satisfy hydraulic, organic, and nitrogenous loading conditions as well as effluent requirements.

Influent to the SBR may be (1) intermittent or single feed, (2) multiple or step feed, or (3) continuous feed. Single feed (Figure 3.4) is most commonly used and provides for best nitrification. Step feed (Figure 3.5) most easily accommodates

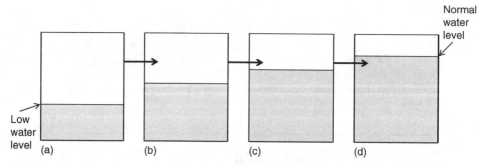

Figure 3.5 *Step-feed SBR. The step feed or multiple feed SBR contains a number of "fill and react" periods (**a**, **b**, and **c**) before the SBR reaches its normal water level (**d**). After reaching the normal water level, the SBR enters an additional React Phase and then Settle Phase and Decant Phase. The step feed SBR may be used for treating high-strength industrial wastewater.*

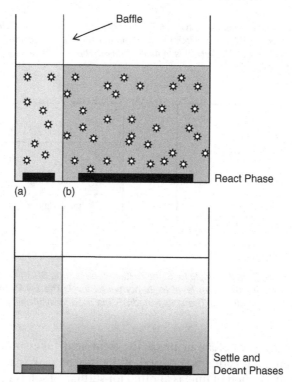

Figure 3.6 *Continuous-fill SBR. In the continuous-flow sequencing batch reactor, influent always enters the reactor. There are two chambers in the reactor that are divided by a baffle. The smaller chamber or per-react zone (a) receives the influent, and from here the influent slowly moves into the larger chamber or main react zone (b). The larger chamber acts as the sequencing batch reactor. However, the sequencing batch reactor only has a limited number of phases: React, Settle, and Decant.*

high strength organic loading conditions. Continuous feed (Figure 3.6) is used infrequently and mostly during periods of high flows that are caused by I/I. Typical phases for each cycle of the single-feed SBR listed sequentially in the treatment process are:

- Fill Phase
- React Phase
- Settle Phase
- Decant Phase

SINGLE FEED: FILL PHASE

During the Fill Phase (Figure 3.4), influent containing substrate or BOD is added to the SBR, raising the liquid level from the minimum level to a depth that corresponds to the amount of influent received during the Fill Phase. The time allocated to the Fill Phase is variable and dependent on the influent flow rate, desired loading condition, detention time, and settling characteristics of the biomass. Typically, SBR are designed to have a minimum fill time that corresponds to the peak hourly flow rate of the treatment plant.

The Fill Phase may be operated in any of the following modes or combination of modes:

- Static Fill (anaerobic/fermentative—no aeration and no mixing) is used to produce volatile fatty acids (Table 3.2) such as formate (HCOOH) and acetate (CH₃COOH) [Eq. (3.1)], which are needed to promote biological phosphorus release (Figure 3.7). Static Fill also is used (1) during start-up of an SBR, (2) to control undesired filamentous organism growth, and (3) during low flow conditions to save electrical costs.
- Mix Fill (anoxic—no aeration but mixing) is used to promote denitrification [Eq. (3.2)] and control undesired growth of filamentous organisms.
- Aerated Fill (oxic—aeration and mixing) is used to reduce cBOD loading before the React Phase and to promote biological phosphorus uptake (Figure 3.8), if Aerated Fill is preceded by Static Fill or biological phosphorus release. Aeration can be provided by most aeration systems, including diffused, floating mechanical, or jet. The aeration system must be able to provide a wide range of mixing intensities and the flexibility of mixing without aeration.

$$\text{cBOD} + \text{organic molecule} \xrightarrow{\text{anaerobic/fermentative condition}} \text{volatile fatty acids} + CO_2 + H_2O \tag{3.1}$$

$$\text{cBOD} + NO_3^- \xrightarrow{\text{anoxic/denitrifying condition}} N_2 + N_2O + OH^- + CO_2 + H_2O \tag{3.2}$$

TABLE 3.2 Examples of Volatile Fatty Acids

Fatty Acid	Formula
Formate	HCOOH
Acetate	CH₃COOH
Propionate	CH₃CH₂COOH
Butyrate	CH₃CH₂CH₂COOH
Valeric acid	CH₃CH₂CH₂CH₂COOH
Caproic acid	CH₃CH₂CH₂CH₂CH₂COOH

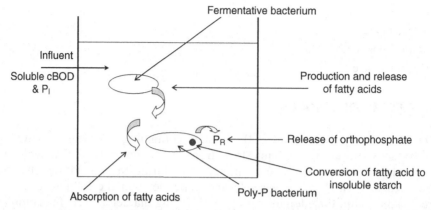

Figure 3.7 *Biological phosphorus release. During biological phosphorus release, an anaerobic/ fermentative condition is established during the Static Fill Phase. In the absence of free molecular oxygen and nitrate, fermentative bacteria or acid-forming bacteria ferment (produce) fatty acids from the influent, soluble cBOD. The acids are released to the bulk solution. Poly-P bacteria absorb the fatty acids and convert them to insoluble starch granules (β-polyhydroxybutyrate). The absorption and conversion of fatty acids to starch granules results in the release of phosphorus to the bulk solution by the Poly-P bacteria. The bulk solution under an anaerobic/fermentative condition contains two "pools" of orthophosphate, the influent orthophosphate (P_I) and the released orthophosphate (P_R).*

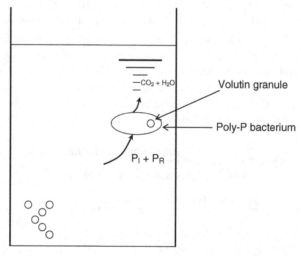

Figure 3.8 *Biological phosphorus uptake. During biological phosphorus uptake, an aerobic condition is established during the React Phase. In the presence of free molecular oxygen, Poly-P bacteria solu- bilize and degrade the starch granules. The degradation of the starch granules results in the release of much energy that is stored by the Poly-P bacteria in the form of polyphosphate or volutin granules. In order to produce the volutin granules, the Poly-P bacteria remove large quantities of orthophosphate (P_I and P_R) from the bulk solution.*

During Mix Fill mixing action blends the influent with the biomass. Without aeration, nitrates are used by facultative anaerobic bacteria or denitrifying bacteria for the degradation of soluble cBOD in the influent. Denitrification occurs in the absence of free molecule oxygen (O_2) or presence of an oxygen gradient. A dissolved oxygen concentration of ≤0.8 mg/L helps to ensure the presence of an oxygen gradient.

During Static Fill and Mix Fill, soluble cBOD is rapidly absorbed by floc bacteria. Filamentous organisms have much difficulty competing with floc bacteria for available soluble cBOD under an anaerobic/fermentative (Static Fill) condition and an anoxic (Mix Fill) condition. Also, strict aerobic filamentous organisms are inactive under nonaerobic conditions. Therefore, Static Fill and Mix Fill contribute to the control of undesired filamentous organism growth as they "select" for the growth of floc bacteria and "select" against the growth of filamentous organisms.

The most important operational condition affecting denitrification is soluble cBOD. The greater the amount of soluble cBOD that is present, the more rapidly residual dissolved oxygen is removed by facultative anaerobic bacteria and the more quickly nitrate is used. Approximately three parts of soluble cBOD or carbon are required for each part of nitrate to be used by facultative anaerobic bacteria.

SINGLE FEED: REACT PHASE

The React Phase begins after the reactor has completed the Fill Phase and the influent has been diverted to another reactor. The React Phase continues until degradation of cBOD is complete and, if necessary, nitrification and/or biological phosphorus uptake also have been completed. Degradation of cBOD usually occurs when the dissolved oxygen concentration in the reactor exceeds 2–3 mg/L, and nitrification usually occurs when the dissolved oxygen concentration in the reactor is approximately 2–3 mg/L.

During the React Phase, no additional influent enters the reactor and aeration and mixing are provided on a continuous or intermittent basis. The purpose of aeration and mixing is to complete cBOD degradation and nitrification and biological phosphorus uptake, if the Static Fill Phase was employed. The purpose of nonaeration during the React Phase is to promote denitrification.

During the React Phase the reactor does not receive substrate, because the influent is discharged to another reactor by an influent control valve. Dissolved oxygen concentration in the reactor increases during continuous aeration and mixing, and the dissolved oxygen concentration cycles between high and low values during intermittent aeration. Intermittent aeration promotes denitrification during the absence of dissolved oxygen or presence of an oxygen gradient. Because no influent or substrate enters the reactor, the rates of cBOD removal and nitrification increase dramatically with increasing time in the React Phase.

SINGLE FEED: SETTLE PHASE

During the Settle Phase, aeration and mixing are terminated. Bacterial solids or floc particles that are responsible for the treatment of the influent are separated

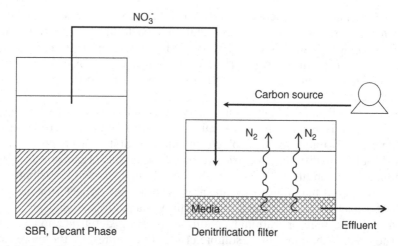

Figure 3.9 *Denitrification tank or filter. If denitrification periods in the SBR are not adequate in reducing the quantity of nitrate in the decant, then the use of a denitrification tank or filter downstream of the SBR is needed. A denitrification filter contains a bed of media (plastic, wood, foam, or sand) that provides for the growth of bacteria (biofilm) that use nitrate when a soluble cBOD source is provided. When nitrate is used in the filter to degrade the carbon source, the nitrogen in the nitrate is released to the atmosphere as molecular nitrogen (N₂).*

from the suspending medium, and a clarified supernatant is produced. The settled solids form a distinctive interface with the clear supernatant. If the solids do not settle and compact properly, some solids can be lost from the reactor during the Decant Phase.

In the sequencing batch reactor the Settle Phase usually is more efficient in obtaining a high-quality supernatant than a continuous flow process, because there is no influent flow to or effluent flow from the reactor; that is, settling occurs in a completely quiescent condition. The entire reactor serves as the clarifier, and there is no need for underflow equipment to return solids to the aeration tank. The loading rate during Settle Phase is zero for the reactor.

If the quantity of nitrate (NO_3^-) remaining in the reactor after React Phase would contribute to a discharge permit violation for nitrate or total nitrogen, then (1) a soluble cBOD or carbon source would be added to the decant in a downstream denitrification tank or filter to remove the nitrates (Figure 3.9) or (2) a carbon source would be added to the settled solids during settle phase to promote denitrification, and a short period of aeration or mixing would be employed after denitrification to "strip" the entrapped gases in the floating solids and then settling of solids would again be permitted (Figure 3.10). If the quantity of nitrates in the mixed liquor following Settle Phase would not contribute to a permit violation, they may be used in a Mix Fill phase to control undesired filamentous organism growth.

A carbon source is needed to promote denitrification, because the quantity of soluble cBOD remaining after React Phase usually is inadequate to promote desired denitrification. Because denitrification results in floating solids due to the captured gases (N_2, N_2O, and CO_2) within the solids, a short period of aeration or mixing is necessary after denitrification to strip the entrapped gases from the solids.

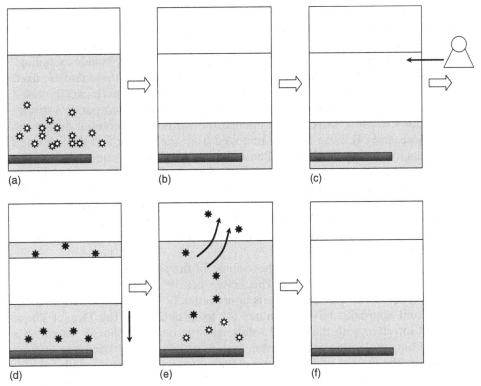

Figure 3.10 *Denitrification after Settle Phase. Denitrification can be achieved in an SBR during Mix Fill Phase and after Settle Phase. To achieve denitrification after Settle Phase, the following events are needed: Nitrification occurs during the React Phase (**a**), Settle Phase occurs (**b**), a soluble carbon source is added to the SBR (**c**), denitrification occurs in the settled solids resulting in the production of molecular nitrogen (N_2) that becomes entrapped in floating solids (**d**), a second and short aeration period is employed to strip the entrapped molecular nitrogen form the floating solids (**e**), and a second settle period occurs that results in the settling of solids (**f**). Following the denitrification period, the SBR enters the Decant Phase.*

SINGLE FEED: DECANT PHASE

During the Decant Phase, clarified or treated supernatant or decant is removed from the sequencing batch reactor. Decanting may be achieved with a pipe fixed at some predetermined level with the flow regulated by an automatic valve or pump or floating or adjustable weir at or beneath the interface between the settled solids and the supernatant. When the liquid level reaches the minimum level the Decant Phase ends.

Floating decanters maintain the inlet orifice slightly below the water surface to minimize the removal of floating solids, foam, and scum during the Decant Phase. Fixed-arm decanters allow the operator to lower or raise the level of the decanter. The decanter volume should be the same as the volume of influent that enters the SBR during the Fill Phase. Also, the vertical distance from the decanter to the bottom of the reactor should be maximized to avoid disturbing the settled solids (sludge blanket).

Early decanters were simply pipes with drilled holes along the bottom and sides. Unfortunately, during Fill Phase, React Phase, and Settle Phase, solids plugged the holes and pipes, resulting in the operational problems of discharging these solids during the beginning of Decant Phase. The development of new "solids excluding" decanters corrected the problem. The new decanters include (1) those that are fixed and air-filled except during the Decant Phase, (2) those that are mechanically closed by a hydraulic or electric motor when not in use, (3) those that incorporate a spring-loaded solids excluding valve that is opened by hydraulic differential during the Decant Phase, and (4) those that are removed from the reactor and then inserted into the reactor during the Decant Phase. Decanters that are removed from the reactor and then inserted into the reactor during the Decant Phase may entrain floating solids, foam, and scum when lowered into the supernatant. Although there are many types of decanting mechanisms, the most commonly used is the floating decanter or adjusting weirs. The decanter withdraws supernatant approximately 18 inches below the liquid surface. The rate of decant is controlled by automatic valves in a gravity system or by pumping.

Typically, approximately 25% of the volume of the sequencing batch reactor is decanted during the Decant Phase. This leaves nearly all of the activated sludge within the reactor. The Decant Phase is approximately 30–45 minutes for floating decanters and approximately 60 minutes for fixed decanters. The Decant Phase should not interfere with the settled solids, and the decanter should not cause a vortex resulting in floating solids. During high flows caused by I/I, influent may be added during the Decant Phase to more quickly receive and treat the influent.

GENERAL DECANTER OPERATIONAL AND DESIGN GUIDELINES

In order to provide for reduced or limited withdrawal of solids from the SBR during the Decant Phase, the following general decanter operational and design guidelines are offered:

- The decanter should draw treated supernatant from below the water surface and exclude foam and scum.
- An adequate zone of separation between the sludge blanket and the decanter should be maintained at all times during the Decant Phase.
- A means for excluding solids from entering the decanter during the React Phase and Settle Phase should be provided.
- Protection against ice buildup on the decanter and freezing of the discharge pipes and decant valves should be provided.

SINGLE FEED: IDLE PERIOD

In a multiple reactor system an Idle Period (Figure 3.11) or "stand-by" time for a sequencing batch reactor may be incorporated in each cycle. This stand-by time is for the reactor just completing its batch cycle while waiting for another reactor to complete its Fill Phase or the time period between the completion of Decant Phase and the beginning of Fill Phase. The length of the Idle Period is determined

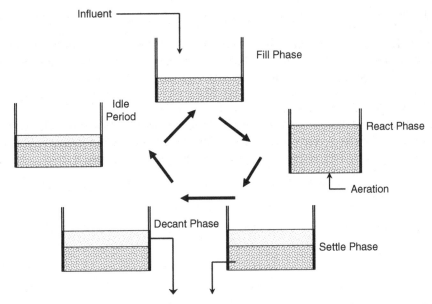

Figure 3.11   *Idle Period. If wastewater flow is relatively low and wastewater strength also is relatively low, then an SBR may have sufficient time after the completion of Decant Phase to "park" the SBR in an Idle Period. During the Idle Period, no significant bacterial activity occurs. The SBR is not fed influent and simply waits for a period of time before it is filled with influent. Although not a true phase of reaction during an SBR cycle, the Idle Period is sometimes referred to as the Idle Phase.*

by the flow rate of the wastewater into the treatment plant. The Idle Period may be eliminated, especially during periods of I/I.

A major advantage of SBR technology as compared to continuous-flow technology is its flexibility to adjust or modify reactor conditions through changes in the length of time of any phase and the dissolved oxygen concentration during Aerated Fill Phase or React Phase. In addition, the length of an aerobic, anoxic, and anaerobic/fermentation phase may be changed to satisfy discharge requirements.

MULTIPLE FEED

Multiple feed or step feed has the identical phases for each cycle of the reactor as single feed, except multiple Fill Phases and multiple React Phases precede the Settle Phase (Figure 3.5), and the sequencing batch reactor is not filled to total fill volume until the completion of the last batch of influent is fed to the SBR.

CONTINUOUS FEED

Wastewater treatment plants that employ sequencing batch reactors commonly use two or more reactors that operate in parallel. However, the reactors may be operated as continuous feed (Figure 3.6). Here, influent enters the reactors on a continuous basis. Some wastewater treatment plants operate SBR under continuous feed as a standard mode of operation, while other wastewater treatment plants operate SBR under continuous feed only during an emergency condition such as high flows caused by I/I. Continuous-feed operation may experience problems such as overflows, washouts, poor effluent quality, and permit violations.

During continuous feed there are only three phases: React Phase, Settle Phase, and Draw Phase. The SBR is divided into two zones by a baffle. These zones are the pre-react zone and main react zone. The pre-react zone serves as a bio-selector that promotes the growth of floc bacteria and limits the growth of filamentous organisms. The pre-react zone also serves as an equalization tank and grease trap.

4

Sludge Wasting

The conversion of biochemical oxygen demand (BOD) to bacterial cells [Eq. (4.1)] and the addition of nondegradable influent solids in the sequencing batch reactor result in the accumulation of living and nonliving sludge or mixed-liquor suspended solids (MLSS). The living portion of the MLSS is referred to as the mixed-liquor volatile suspended solids (MLVSS). If MLSS are not removed or wasted from the SBR in proper quantity and timing, much of the MLSS would be loss in the reactor decant.

$$1 \text{ pound domestic or municipal BOD} + O_2 \rightarrow 0.6 \text{ pound sludge (MLVSS)} \quad (4.1)$$

There are several operational conditions that influence the quantity of sludge produced in the SBR (Table 4.1). These factors include (1) mean cell residence time (MCRT), (2) food-to-microorganism (F/M) ratio, and (3) organic loading.

Mean cell residence time is commonly used at activated sludge processes (including SBR) as a guideline to provide for an adequate number of bacteria (MLVSS) for efficient treatment of influent. With increasing MCRT, an increasing number of bacteria are present in the SBR; conversely, with decreasing MCRT, a decreasing number of bacteria are present in the SBR. Changes in MCRT or sludge age through sludge wasting affect the following operational conditions:

- Dissolved oxygen concentration
- Growth rate of the bacteria
- Nutrient requirements
- Settleability of the MLSS
- Decant quality

Troubleshooting the Sequencing Batch Reactor, by Michael H. Gerardi
Copyright © 2010 by John Wiley & Sons, Inc.

TABLE 4.1   Operational Conditions that Influence Sludge Production in an SBR

Operational Condition	Results in Decreased Sludge Production
MCRT	High values
F/M	Low values
Organic loading	Low loadings
TSS-to-BOD ratio	Low ratios
Use of O_2 or NO_3^- to degrade soluble BOD	Use of NO_3^-

Wasting of MLSS may occur immediately after the React Phase or during the Settle Phase, the Decant Phase, or the Idle Period. Sludge wasting usually occurs at the end of the Decant Phase when sludge compaction is greatest. By wasting at the end of the Decant Phase, the highest concentration of solids is removed, and therefore the lowest volume of wastewater is removed from the reactor. Multiple discharge points provide for the thickest solids. Sludge wasting is not a phase of operation for an SBR.

If a submersible pump is used for wasting solids from the SBR, and the pump is mounted in a corner of the SBR, care should be exercised that a "rat hole" does not develop. A rate hole occurs when the submersible pump sucks a hole in the sludge blanket, and the majority of the wasted liquid is water.

In sequencing batch reactors, only occasional wasting is required. Small frequent adjustments of MLSS through sludge wasting are better than large infrequent adjustments. Frequent wasting of solids is recommended to maintain the desired age, biomass health, and solids settling characteristics. Wasting is determined by the desired MCRT for the SBR. Sludge is removed from the reactor by decreasing the volume of mixed liquor and decreasing the quantity of MLSS in the reactor. The adjustment of MLSS inventory is performed by checking the MLSS or MLVSS concentration and the settled sludge interface level.

As compared to the conventional, activated sludge process, there is no underflow, no recycle solids, and no return pumps in the sequencing batch reactor. This simplifies the operational requirements for solids wasting. The operator needs only to check the MLSS or MLVSS inventory and the settled sludge interface with a sludge measuring device or sludge judge. The sludge blanket should be a safe distance below the bottom water level at the end of the Decant Phase.

The maintenance of a constant MCRT is not as important in the sequencing batch reactor as the MCRT is in the conventional, activated sludge process. However, changes in sludge age should be made slowly. Higher sludge ages do favor nitrification and the undesired growth of some filamentous organisms.

Part II

Substrate

5

BOD

Substrate is the food or carbon and energy source for bacteria. The quantity of substrate present in a waste stream is typically measured as biochemical oxygen demand (BOD) or chemical oxygen demand (COD).

BOD is measured as mg/L and is one of the most widely used analytical tests to measure the strength of wastewater. The design of wastewater treatment processes, including activated sludge, trickling filter, and ponds, are often determined in large part by using the BOD of the influent. The size of aeration equipment, reactor capacity, and pond surface normally are based upon the influent BOD.

The BOD test was developed in the United Kingdom at the beginning of the twentieth century. The test was based upon the assumption that in polluted water there is a demand for oxygen caused by microorganisms, primarily bacteria, and this demand could be used to measure the amount of pollution, that is, the greater the demand for oxygen, the greater the pollution.

The BOD test was designed as a 5-day test. The duration of the test was based on the assumption that most surface waters in the United Kingdom take 5 days or less to drain to the sea. Several countries have modified the duration of the test over the years. There are 5-day BOD (BOD_5), 7-day BOD (BOD_7), and 21-day BOD (BOD_{21}) tests, and there is an ultimate BOD (BOD_u) test.

The ultimate BOD value is the BOD that would be exerted at infinite time (approximately 100 plus days). Because the BOD_u is difficult and time-consuming to measure, it is usually determined by constructing a curve of the BOD versus time (Figure 5.1).

Historically, the biochemical oxygen demand as applied to wastewater is the 5-day BOD. The BOD_5 was recommended and adopted for use by The Royal Commission on Sewage Disposal in 1908. This analytical test involves the measurement of the quantity of dissolved oxygen used by microorganisms, primarily

Troubleshooting the Sequencing Batch Reactor, by Michael H. Gerardi
Copyright © 2010 by John Wiley & Sons, Inc.

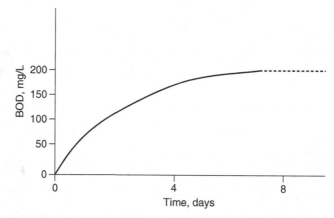

Figure 5.1 *Ultimate BOD. Although BOD$_5$, BOD$_7$, and BOD$_{21}$ tests are commonly used to address the strength (pollution value) of a waste stream, the ultimate demand for oxygen to degrade the pollutants is determined by the ultimate BOD test (BOD$_u$). With increasing time, the amount of dissolved oxygen removed from the bulk solution by microorganisms to degrade the pollutants gradually decreases and the slope of the line (BOD, mg/L versus time, days) begins to parallel the x axis as noted by the change from solid line to breaks in the line.*

bacteria, in their biochemical oxidation of organic or carbonaceous matter. Therefore, the BOD or BOD$_5$ is at times referred to as the carbonaceous, biochemical oxygen demand (cBOD) of the wastewater.

The oxygen demand associated with the oxidation of ammonium (NH_4^+) to nitrate (NO_3^-) is called the nitrogenous biochemical oxygen demand (nBOD). Nitrifying bacteria perform the oxidation of ammonium to nitrate. Because the growth rate or generation time of nitrifying bacteria is high, typically 3–6 days in wastewater, it usually takes several days for a significant population of nitrifying bacteria to develop and exert a measurable oxygen demand in the BOD or BOD$_5$ test. If the nitrifying bacterial population is not present in significant numbers, then the BOD or BOD$_5$ test is a measurement of the cBOD.

However, if sufficient numbers of nitrifying bacteria are present initially or are introduced with the "seed" that is used in the BOD test, the oxygen demand exerted by the nitrifying bacteria can be significant. Therefore, the BOD or BOD$_5$ would be a measurement of carbonaceous and nitrogenous demands. To inhibit the oxygen demand exerted by a nitrifying population, an inhibitory compound such as thiourea ((NH_2)$_2$CS) or allylthiourea ($C_3H_5NHCSNH_2$) is added to the BOD bottles that are used in the test. With the addition of an inhibitory compound, the resulting oxygen demand as determined by testing is the carbonaceous demand or cBOD only. However, the addition of a nitrification inhibitor also can adversely affect the respiration rate of some organotrophic bacteria or cBOD-removing bacteria. This would reduce the total mass of oxygen consumed in a BOD test, if this were to occur.

If nitrifying bacteria exert a significant oxygen demand in the BOD test, then the BOD value used in calculating the oxygen demand for organic or carbonaceous loading is incorrect. A nitrogenous demand has been included, and the oxygen demand for carbonaceous waste is calculated higher than the actual demand in the wastewater.

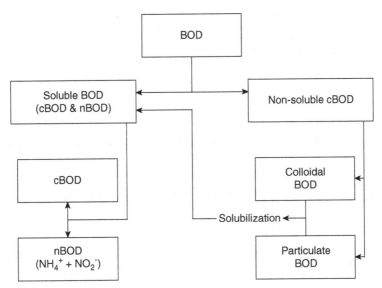

Figure 5.2 *Forms of carbonaceous wastes or cBOD. There are several forms of carbonaceous wastes or cBOD that make up BOD (total BOD or BOD$_5$). These forms are first divided into soluble BOD and non-soluble BOD. Soluble BOD consists mostly of carbonaceous forms that are simple in structure and include short-chain acids, alcohols, sugars, and amino acids and two nitrogenous forms of BOD (nBOD). These nitrogenous forms are ammonium (NH$_4^+$) and nitrite (NO$_2^-$). Nonsoluble forms of cBOD include colloidal BOD such as proteins and particulate BOD such as starches. If sufficient time is provided in an SBR or additional treatment tanks such as a digester, the colloidal BOD and particulate BOD can be solubilized to simple acids, alcohols, sugars, and amino acids and then degraded.*

It is common practice in many wastewater laboratories to routinely calculate the cBOD with the use of a nitrifying inhibitor and occasionally calculate the BOD without the use of a nitrifying inhibitor. Therefore, to better determine the oxygen demand of a wastewater, a BOD test and a cBOD test should be performed.

Carbonaceous wastes that exert a cBOD demand may be soluble, colloidal, or particulate (Figure 5.2). Soluble cBOD consists of compounds that are dissolved in wastewater and are <0.1 µm in size. Acetate (CH$_3$COOH), ethanol (CH$_3$CH$_2$OH), and glucose (C$_6$H$_{12}$O$_6$) are examples of soluble cBOD. Colloidal cBOD consists of compounds with a relatively large surface area that are not dissolved in wastewater and are suspended in wastewater. Compounds that are colloidal cBOD are approximately 0.1–1.0 µm in size. Proteins are examples of colloidal cBOD. Particulate cBOD consists of insoluble, microscopic and macroscopic wastes such as cellulose (vegetable scrapes) and lignin (wood fibers).

Nitrogenous BOD or nBOD consists of ammonium and nitrite (NO$_2^-$). In order to obtain energy, nitrifying bacteria oxidize ammonium to nitrite [Eq. (5.1)] and then nitrite to nitrate [Eq. (5.2)]. Ammonium is found in wastewater mostly through the hydrolysis of urea (NH$_2$COH$_2$N), a major component of urine [Eq. (5.3)]. Unless an industry discharges nitrite (Table 5.1), nitrite is not found in the raw wastewater. Ammonium can be released in an SBR through the ammonification of organic-nitrogen compounds such as proteins and amino acids or the die-off of large numbers

TABLE 5.1   Nitrite-Containing Industrial Wastewater

Corrosion inhibitors
Leachate (biologically pretreated)
Meat processing (preservatives or pretreated)
Steel mill

of bacteria. Ammonification [Eq. (5.4)] results in the release of amino groups (-NH$_2$) from organic-nitrogen compounds and the production of ammonium.

$$NH_4^+ + 1.5O_2 \xrightarrow{\text{Ammonium-oxidizing bacteria}} NO_2^- + 2H^+ + H_2O \quad (5.1)$$

$$NO_2^- + 0.5O_2 \xrightarrow{\text{Nitrite-oxidizing bacteria}} NO_3^- \quad (5.2)$$

$$NH_2COH_2N + H_2O \xrightarrow{\text{Hydrolytic bacteria}} 2NH_3 + CO_2 \quad (5.3)$$

$$\text{Amino acid} \xrightarrow{\text{Ammonification}} NH_4^+ + \text{organic acid} \quad (5.4)$$

The quantity of ammonium that can be released in an SBR through ammonification can be determined by testing (1) the influent ammonia (NH$_3$) and (2) the influent total Kjeldahl nitrogen (TKN). The TKN measures the quantity of influent ammonia and the quantity of ammonia that is produced from the release of amino groups from organic-nitrogen compounds. Therefore, the difference between the influent TKN and the influent ammonia is the ammonia that is available from the organic-nitrogen compounds. The influent TKN value is always greater than or equal to the influent ammonia value.

Ammonia (NH$_3$) and ammonium (NH$_4^+$) are reduced forms of nitrogen, that is, they contain hydrogen. Although both forms of reduced nitrogen are present in wastewater, ammonium is the dominant form at pH values of <9.4 while ammonia is the dominant form at pH values of ≤9.4 (Figure 5.3).

The quantity of ammonium in wastewater at near-neutral pH values (6.8–7.2) is determined by measuring the quantity of ammonia in the wastewater or stripped from the wastewater after the pH of the wastewater has been increased to >9.4. With the increase in pH, ammonium is converted to ammonia. Ammonia is easily stripped from the wastewater when the wastewater is stirred, and the stripped ammonia is then measured by an ammonia probe. Therefore, by measuring the ammonia directly that is stripped from the wastewater, ammonium is measured indirectly.

BOD analyses require much care and quality control. Significant errors can result in inaccurate and misleading BOD values that affect the operation and efficiency of the treatment process and surcharge billings for excess BOD. Over-dechlorination of a wastewater sample with sodium thiosulfate (Na$_2$S$_2$O$_3$) increases the resulting BOD value as sodium thiosulfate removes dissolved oxygen from the BOD water. Toxicity may or may not affect BOD analyses.

If toxic waste is diluted below its toxic concentration when it is introduced into the BOD bottle, no significant change in the expected or typical value of BOD for

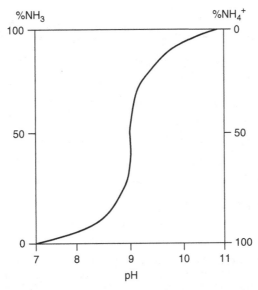

Figure 5.3 *Relative abundance of reduced nitrogen (NH₃ and NH₄⁺). The relative abundance or amount of reduced nitrogen as ammonia (NH₃) and ammonium (NH₄⁺) is dependent upon the pH of the SBR. With increasing pH, ammonia increases in quantity and conversely with decreasing pH ammonium increases in quantity. Shown here is the change in the dominant form of reduced nitrogen at pH values of <9.4 (ammonium favored) and of ≥9.4 (ammonia favored).*

the wastewater occurs. If the toxic waste kills a large number of bacteria in the BOD bottle, then dissolved oxygen consumption for cellular respiration and degradation of wastes is depressed and a significant decrease in the expected or typical value of BOD for the wastewater occurs. If the toxic waste kills a small or specific group of bacteria, the dead bacteria then serve as an additional substrate for the surviving bacteria. The additional substrate results in more dissolved oxygen consumption by the surviving bacteria as these bacteria degrade the dead bacteria. Therefore, a significant increase in the expected or typical value of BOD of the wastewater occurs.

When seed is used to introduce bacteria into the BOD test, it is assumed that the bacterial population is acclimated to the substrate (BOD) being tested. This usually is true. However, if the standard seed is used for an industrial wastewater or a wastewater sample that contains compounds that are degradable by a specific group of bacteria, the bacteria are probably not acclimated to the substrate. Under these conditions the BOD measurement would be false, because the bacteria in the seed may take a relatively large amount of time over the 5-day BOD test to acclimate and grow an adequate population to degrade the substrate.

Also, there are oxidizable wastes such as sulfides (HS⁻) in many waste streams that will be oxidized chemically in a BOD bottle without bacterial activity. The chemical oxidation will contribute to a false and elevated BOD value.

The BOD test does not tell the operator anything about the rate of degradability of the waste or its ultimate strength (Figure 5.4). This may mislead the operator with respect to the impact of the BOD on the treatment process.

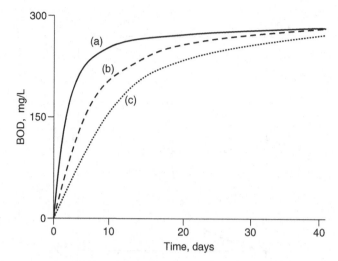

Figure 5.4 *Rate of BOD degradation. Although three different BOD (**a**, **b** and **c**) may demand approximately the same quantity of oxygen for their degradation over 20 or 30 days, their rate of degradation over five days may be significantly different, where one BOD (**a**) may remove a large quantity of oxygen from the SBR over a short period of time, while another BOD (**c**) may remove a small quantity of oxygen from the SBR over the same period of time.*

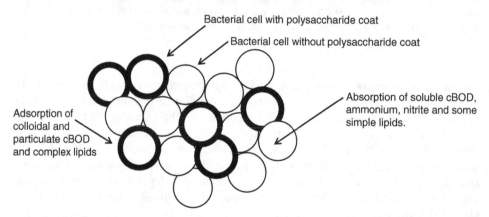

Figure 5.5 *Absorption and adsorption of BOD. Not all forms of BOD are easily and rapidly degraded. Only soluble cBOD such as simplistic acids, alcohols, sugars and amino acids are able to pass through the cell wall and cell membrane of bacterial cells, where they are then degraded by the cells. Some simple lipids also are able to pass through the cell wall and cell membrane of bacterial cells. Soluble nBOD (ammonium and nitrite) also are able to pass through the cell wall and cell membrane of nitrifying bacteria, where they are degraded. Colloidal and particulate cBOD must first come in "contact" with the polysaccharide coating of bacterial cells, where they are solubilized over time. Solubilization of the colloidal and particulate cBOD results in the production of soluble acids, alcohols, sugars, and amino acids that are then absorbed by bacterial cells that have and do not have the polysaccharide coating. Once absorbed, these compounds are then degraded; that is, they are stabilized. The adsorption and then absorption/degradation of BOD is known as "contact-stabilization."*

In order for BOD to be degraded in an SBR, it must be absorbed by bacteria (Figure 5.5). Therefore, soluble and simplistic BOD is degraded first. Insoluble and complex BOD must be solubilized or converted by bacteria into soluble and simplistic BOD in order to be absorbed and degraded. This requires time (Table 5.2).

TABLE 5.2   Approximate MCRT Required for the Beginning of BOD Degradation

Biological Event	Approximate MCRT Required
Degradation of soluble cBOD	0.3 days (7 hours)
Solubilization of colloidal cBOD	2 days
Solubilization of particulate cBOD	2 days
Degradation of xenobiotic cBOD	5 days
Nitrification (degradation of nBOD)	3–15 days (temperature-dependent)

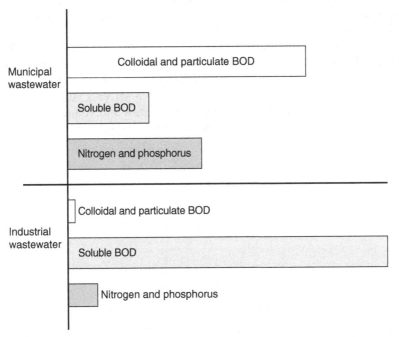

Figure 5.6 *Comparison of major components of municipal and industrial wastewaters. Major components of municipal and industrial wastewaters with respect to the degradation of BOD consist of the types of BOD (colloidal, particulate, and soluble) and their relative abundance as well as the relative abundance of readily available nutrients (ammonical-nitrogen and orthophosphate-phosphorus). Municipal wastewater (top half of the bar graph) contains much colloidal and particulate BOD that degrades slowly; that is, it must undergo solubilization first before it can be absorbed and degraded. This mode of bacterial action does not place an immediate and significant demand on the SBR for dissolved oxygen and nutrients. In addition, municipal wastewater typically has a relatively large quantity of readily available nitrogen and phosphorus for bacterial use. With few exceptions, industrial wastewater has a relatively small quantity of colloidal and particulate BOD and a relatively large quantity of soluble BOD (bottom half of the bar graph). This relatively large quantity of easily and rapidly degradable cBOD places an immediate and large demand on the SBR for dissolved oxygen and readily available nutrients. Unfortunately, most industrial wastewaters are deficiency for readily available nitrogen (ammonical-nitrogen) and/or phosphorus (orthophosphate-phosphorus).*

Therefore, as the quantity of BOD remaining in the SBR decreases, the quantity of BOD is progressively more difficult to degrade.

The difference in the rate of degradation of different types of BOD can be seen between municipal and industrial wastewater (Figure 5.6). Most of the BOD in municipal wastewater consists of colloidal and particulate BOD that degrade slowly.

With some exceptions, most industrial wastewater consists of soluble BOD. When soluble BOD enters the SBR, bacteria rapidly absorb it within 10–15 minutes.

Organotrophic bacteria in an aerobic environment use organic compounds for energy and growth (carbon assimilation). The organic compounds are used as an energy substrate and carbon source for growth and reproduction. Carbon not used by the cell for structural growth is released from the cell as carbon dioxide (CO_2). The ratio of the mass of carbon that is used for growth versus that used for energy is called the organism yield, typically <1. For the BOD test, only the carbon used as an energy source exerts a BOD. For typical domestic or municipal wastewater without significant industrial wastes, the organism yield or sludge yield is 0.67 pound per pound of cBOD.

The degradation of BOD results in oxygen consumption or utilization. The rate of oxygen utilization or respiration rate (RR) is expressed as milligrams of oxygen used per hour per gram of mixed liquor volatile suspended solids (mg O_2/hr/g MLVSS). For an SBR the RR can be 8–27 mg O_2/hr/g MLVSS. Higher RR are associated with higher food-to-microorganism (F/M) ratios and younger sludge. Lower RR are associated with lower F/M ratios and older sludge.

6

COD

The chemical oxygen demand (COD) test is a measurement of the oxygen demand (mg/L) exerted by chemical oxidation of the organic compounds in wastewater. The basis for the COD test is that all organic compounds can be completely oxidized to carbon dioxide (CO_2) with a strong oxidizing agent under acidic conditions. The COD test is used more often than the BOD test to characterize the strength of industrial wastewater.

The potassium dichromate ($K_2Cr_2O_7$) reflux technique is most commonly used for COD testing. Potassium dichromate is a strong oxidizing agent under an acidic condition. Acidity usually is achieved in the COD test with the addition of sulfuric acid (H_2SO_4). As such, it is preferred over COD procedures using other oxidizing agents such as potassium permanganate ($KMnO_4$). It is the most effective COD test, is relatively inexpensive, and is able to almost completely (>95% of the theoretical value) oxidize all organic compounds. Volatile organic compounds and aromatic hydrocarbons are not oxidized during the COD test. For example, benzene (C_6H_6) (Figure 6.1), an aromatic hydrocarbon, is a highly biodegradable compound that is poorly oxidized in the COD test. Therefore, the BOD value is greater than the COD value for benzene. The dichromate reflux technique does not oxidize ammonium to nitrate, so nitrification demand for oxygen is not a component of the COD test.

When organic compounds are oxidized during the COD test, potassium dichromate is reduced; that is, hexavalent chromium, Cr^{6+}, is reduced to trivalent chromium, Cr^{3+}. The amount of Cr^{3+} that is produced is determined, and that amount is used as an indirect measure of the organic content or strength of the wastewater, that is, the more Cr^{3+} that is produced, the greater the strength of the wastewater or pollution.

The COD test is easier to measure and provides a more rapid indicator of the strength of the wastewater than the BOD test. The COD test also indicates the

Troubleshooting the Sequencing Batch Reactor, by Michael H. Gerardi
Copyright © 2010 by John Wiley & Sons, Inc.

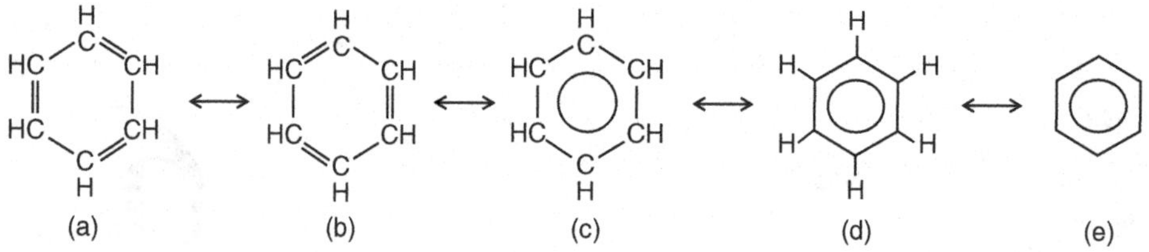

Figure 6.1 Benzene. Benzene is the basic structure for aromatic compounds. As an aromatic compound, benzene is a six-sided figure containing six carbon atoms joined by three single bonds (–) and three double bonds (=) (**a**). However, the pair of electrons that make up the double bonds shift between the carbon atoms resulting in changes in locations of the double bonds (**b**). To better represent the changing positions of the double bonds, the six-sided benzene structure is often represented as a ring of single bonds between the carbon atoms that surround a circle (**c**). Sometimes the carbon atoms are not shown in the benzene structure (**d**), and sometimes the benzene structure is simply illustrated as a six-sided figure that surrounds a circle (**e**).

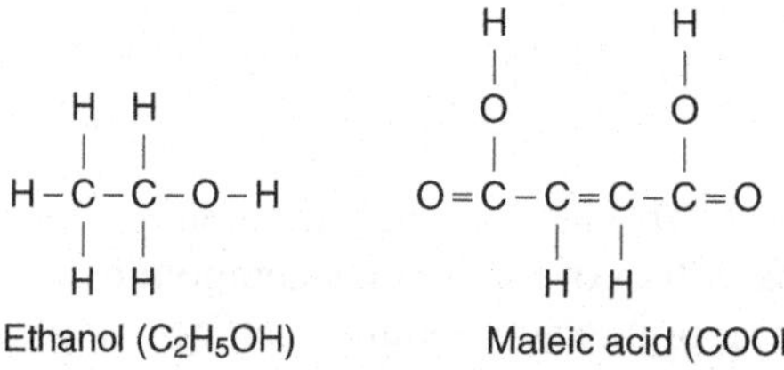

Ethanol (C₂H₅OH)   Maleic acid (COOHCHCHCOOH)

Figure 6.2 Molecular structures and biodegradability. The simpler the chemical structure of an organic compound (cBOD), the more easily the compound is degraded. Ethanol (C_2H_5OH) is a relatively simple organic compound with only single bonds between each atom in the molecule. Aside from the hydroxyl groups (-OH) attached to the molecule, there is no significant degree of substitution of special groups on the carbon chain of atoms. With an increasing number of double bonds, along with increasing substitutions of special groups such as hydroxyl (-OH), carboxyl (-COOH), and branches in the chain of carbon, degradation of the organic compound becomes more difficult, that is, it takes more time and often requires bacteria with specialized enzyme systems that can remove the special groups and break double bonds. Maleic acid (COOHCHCHCOOH) contains two double bonds and two carboxyl groups.

organic concentration in the presence of toxic compounds that may cause interference in the BOD test, that is, the COD test is not inhibited by toxic compounds. However, COD analysis measures biologically degradable and non-biologically degradable organic compounds. Therefore, adjustments in the COD value must be made to define these organic compounds; that is, the COD value is greater than the BOD value of a wastewater. Guideline examples of molecular structures and biodegradability are present in Figure 6.2, and their BOD-to-COD ratios are presented in Table 6.1.

Chemicals that contain carbon and hydrogen are organic compounds. These compounds are divided into two groups, aliphatic and aromatic. The carbon units in aliphatic compounds may be joined in straight chains such as butanol, branch chains such as isobutanol, or cyclic compounds (nonaromatic rings or alicyclic) such as cyclohexane (Figure 6.3). The carbon units may be joined or bonded by a single bond, double bond or triple bond (Figure 6.4).

TABLE 6.1   Examples of Molecular Structure and Biodegradability

Chemical	Formula	Structure	BOD-to-COD	Biodegradability
Acetone	$OC(CH_3)_2$	Aliphatic	0.4	Readily
Benzaldehyde	C_6H_5CHO	Aromatic	0.8	Slowly
Benzoic acid	C_6H_5COOH	Aromatic	0.7	Slowly
Ethyl acetate	$CH_3COOCH_2CH_3$	Aliphatic	0.8	Readily
Ethylene glycol	$HOCH_2CH_2OH$	Aliphatic	0.3	Readily
Maleic acid	$HOCC=CCOOH$	Aliphatic	0.8	Slowly
Methanol	CH_3OH	Aliphatic	0.9	Readily
Toluene	$C_6H_5CH_3$	Aromatic	0.6	Slowly

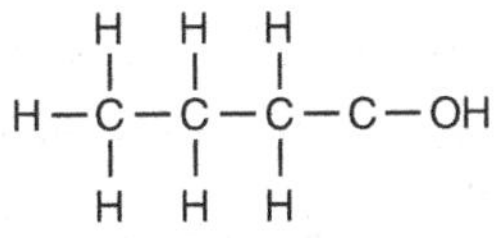

n-Butanol (C_4H_7OH)

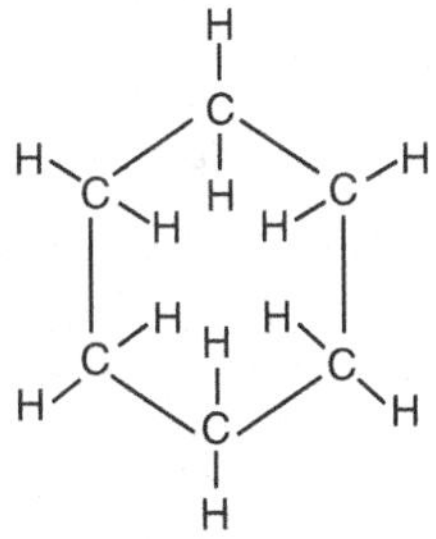

Cyclohexane (C_6H_{12})

Figure 6.3 *Straight-chain and cyclic aliphatic compounds. Straight-chain and cyclic aliphatic compounds (cBOD) consist of a row or rows (branched) of carbon atoms or a cyclic arrangement of carbon atoms. These compounds may possess double-bonded atoms. Straight-chain compounds such as n-butanol (C_4H_7OH) are easily degradable. Cyclic aliphatic compounds such as cyclohexane are also easily degradable, but due to the cyclic construction of the molecule it degrades more slowly than n-butanol.*

Aromatic compounds usually have enhanced chemical stability as compared to similar nonaromatic compounds and therefore are more slowly biodegradable than aliphatic compounds. The benzene ring (Figure 6.1) is the basic aromatic structure. Because the double bonds in the ring shift between carbon units, the benzene ring often is illustrated as a six-sided ring surrounding a circle (Figure 6.5).

There are no firm ratios for BOD-to-COD. Each treatment process must develop its own acceptable ratios for process control and troubleshooting purposes. A typical BOD-to-COD ratio for domestic wastewater is approximately 0.4, while a typical ratio for municipal wastewater is approximately 0.6.

Since only biodegradable organic compounds are removed in the sequencing batch reactor, the COD remaining in the supernatant of a properly operated sequencing batch reactor during the Decant Phase consists mostly of (1) nondegradable organics and (2) soluble, nondegradable microbial waste products.

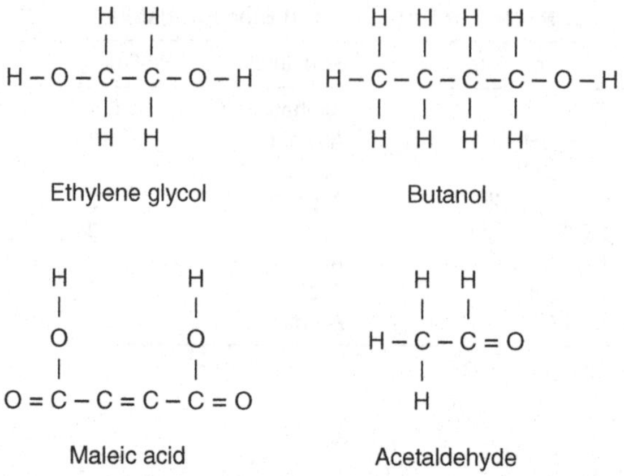

```
      H  H                    H   H   H   H
      |  |                    |   |   |   |
H – O – C – C – O – H    H – C – C – C – C – O – H
      |  |                    |   |   |   |
      H  H                    H   H   H   H
```

Ethylene glycol   Butanol

```
      H          H              H   H
      |          |              |   |
      O          O         H – C – C = O
      |          |              |
O = C – C = C – C = O           H
```

Maleic acid   Acetaldehyde

Figure 6.4 *Single bond and double bonds in organic compounds. Compounds or cBOD that have only single bonds within their molecular structure such as ethylene glycol ($HOCH_2CH_2OH$) and butanol ($CH_3CH_2CH_2CH_2OH$) are more easily and quickly degraded than compounds or cBOD that have double bonds within their molecular structure such as maleic acid (HOOCCCCOOH) and acetaldehyde (CH_3CHO).*

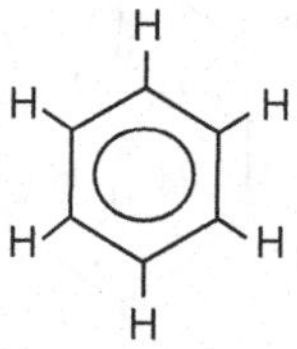

Figure 6.5 *Basic aromatic structure. The basic structure for all aromatic compounds is based on benzene (C_6H_6).*

Because BOD is a biological test and COD is a chemical test, there is no linear relationship. In the BOD test, some of the substrate (organic compounds) is converted to biomass (new cells or sludge), and some of the substrate is oxidized (respired) and contributes to dissolved oxygen depletion and an increase in the BOD value. The organic compounds that are converted to biomass do not contribute to dissolved oxygen depletion and an increase in BOD value. The organic compounds that are respired contribute to oxygen depletion and an increase in BOD value. However, parallel testing for COD and BOD may be beneficial, because COD testing can be used to target a specific BOD range. The BOD-to-COD ratio may be used to determine the quantity of wastewater to be placed in BOD bottles and may reduce multiple BOD dilutions.

INTERFERENCE

High levels of oxidizable inorganic wastes in wastewater samples may interfere with the determination of the COD value. Some inorganic wastes that cause interference include chloride (Cl^-), nitrite (NO_2^-), ferrous iron (Fe^{2+}), and sulfide (HS^-). Chloride is one of the most significant wastes of interference due to its reaction with hexavalent chromium (Cr^{6+}) [Eq. (6.1)]:

$$6Cl^- + Cr_2O_7 + 14H^+ \rightarrow 3Cl^2 + Cr^{3+} + 7H_2O \qquad (6.1)$$

where Cr in Cr_2O_7 has a valence or charge of +6.

Mercuric sulfate ($HgSO_4$) is added to the COD test to remove complex chloride ions present in the wastewater. Without mercuric sulfate the chloride ions would form chloride compounds, and these compounds (not dichromate) would oxidize the organic compounds in the wastewater resulting in a COD value lower than the actual value. Silver sulfate ($AgSO_4$) is added as a catalyst for the oxidation of short, aliphatic (chain) alcohols. Without silver sulfate, dichromate would not oxidize the alcohols, and the COD value of the wastewater would be lower than the actual value. Sulfamic acid (H_3NSO_3) is added to the COD test to remove interference caused by nitrite. Without sulfamic acid the COD value of the wastewater would be higher than the actual value.

BOD-to-COD ratios give the fraction of matter that is biodegradable. There are three fairly reliable guidelines for the correlation (ratios) between BOD and COD. These correlations are:

0.4–0.25 for domestic wastewater

0.6–0.5 for municipal wastewater

0.25–0.17 for industrial wastewater

≤0.01 for relatively nonbiodegradable wastewater

If a consistent ratio—for example, 0.67—is obtained for municipal wastewater, then a conversion with a reasonable degree of confidence can be used to determine the BOD value from the COD value, for example, BOD = 0.67 × COD.

Caution must be exercised when low BOD values are obtained. These values may be reflective of the presence of toxicity or the presence of organic compounds that required more than five days for bacteria to degrade the compounds. Slowly degrading compounds require an ultimate BOD test to determine their impact upon a treatment process.

Part III

Troubleshooting Keys

7

Introduction to Troubleshooting Keys

Troubleshooting keys are provided to assist operators in the identification of the operational condition or conditions that are responsible for the inability of the treatment process to (1) satisfy a discharge permit requirement, (2) correct an operational upset and (3) operate the process in a cost-effective mode. As regulatory agencies place ammonia, total nitrogen, and total phosphorus requirements on more wastewater treatment facilities and as these requirements as well as those for BOD and TSS become more stringent, treatment facilities must maintain effective process control and troubleshooting measures. Part III presents seven chapters that provide reviews, "keys," and tables to assist operators in addressing the following critical operational conditions:

- Nitrification
- Denitrification
- High decant BOD
- High decant TSS
- Undesired change in alkalinity and pH
- Foam and scum production
- Low dissolved oxygen

The chapters provide background information for each of the operational conditions, including specific factors that influence the ability of the treatment facility to (1) successfully satisfy a discharge permit, (2) maintain adequate alkalinity, proper pH, and appropriate dissolved oxygen, and (3) reduce the production and accumulation of foam and scum. Each chapter has a "troubleshooting key" that is used by

Troubleshooting the Sequencing Batch Reactor, by Michael H. Gerardi
Copyright © 2010 by John Wiley & Sons, Inc.

answering a series of numbered questions. The operator's answer to each question leads to another question or the operational factor that is responsible entirely or in part for the problematic condition. Following each troubleshooting key is a "check list" or table of the operational factors to be monitored. The operator can quickly identify what factors have and have not been monitored in the treatment facility by using the table.

8

Troubleshooting Nitrification

BACKGROUND

Nitrification is the conversion or oxidation (addition of oxygen) to ammonium (NH_4^+) to nitrite (NO_2^-) and/or the conversion or oxidation of nitrite to nitrate (NO_3^-). Nitrification is a strictly aerobic (oxygen dependent) reaction that consumes large quantities of dissolved oxygen. Nitrification is performed by two groups of nitrifying bacteria.

The first group of bacteria is the ammonia-oxidizing bacteria (AOB) and consists mostly of *Nitrosomonas* and *Nitrosospira*. This group can only oxidize ammonium to nitrite [Eq. (8.1)]. The second group of bacteria is the nitrite-oxidizing bacteria (NOB) and consists mostly of *Nitrobacter* and *Nitrospira*. This group can only oxidize nitrite to nitrate [Eq. (8.2)].

$$NH_4^+ + 1.5O_2 \xrightarrow{\text{AOB}} NO_2^- + 2H^+ + H_2O \tag{8.1}$$

$$NO_2^- + 0.5O_2 \xrightarrow{\text{NOB}} NO_3^- \tag{8.2}$$

Ammonium (NH_4^+) and ammonia (NH_3) are reduced forms of nitrogen; that is, they contain hydrogen. Ammonium is present in wastewater in large quantities at pH values of <9.4, while ammonia is present in wastewater in large quantities at pH values of ≥9.4. When testing and reporting the concentration of reduced nitrogen in a wastewater sample, the analytical technique requires that the pH of the sample be increased with an alkali compound to a pH of >9.4. This increase in pH converts ammonium in the wastewater to ammonia that is either measured by a probe in the wastewater by a probe or stripped from the wastewater by mixing action and

---

Troubleshooting the Sequencing Batch Reactor, by Michael H. Gerardi
Copyright © 2010 by John Wiley & Sons, Inc.

TABLE 8.1 Industrial Dischargers of Ammonium, Nitrite, and Nitrate

Industry	Ammonium	Nitrite/Nitrate
Automotive	X	
Chemical	X	
Coal	X	
Corrosion inhibitor		X
Fertilizer	X	
Food		X
Leachate	X	
Leachate, pretreated		X
Livestock	X	
Meat	X	
Meat, preservative		X
Ordnance	X	
Petrochemical	X	
Pharmaceutical	X	
Primary metal	X	
Refineries	X	
Steel		X
Tanneries	X	

measured by a probe in the atmosphere immediately above the wastewater. By testing for ammonia, the concentration of ammonium is indirectly measured.

Ammonia is toxic and serves no useful biological purpose. Ammonium serves two significant biological roles. First, it is the primary nitrogen nutrient for bacterial growth (sludge production). Second, it serves as the energy substrate for nitrifying bacteria. Nitrifying bacteria oxidize ammonium and nitrite only to obtain energy for cellular activity and growth. Approximately 60% to 80% of the nitrogen in domestic and municipal wastewaters is in the form of ammonium. The remaining nitrogen is in the organic-nitrogen form.

The degree of nitrification or nitrification/denitrification, if required, that must be achieved by an SBR is determined by the maximum allowable limit of ammonia or total nitrogen, respectively, that is listed on the treatment plant's discharge permit. Total nitrogen consists of nitrite, nitrate, and total Kjeldahl nitrogen (TKN). TKN consists of organic nitrogen and ammonia. The quantity and quality of nitrogenous wastes in the influent of municipal wastewater treatment process may change significantly, depending on the quantity and quality of industrial discharges (Table 8.1).

An SBR nitrifies for three reasons. These reasons are (1) it is required by a regulatory agency to satisfy an ammonia or total nitrogen discharge limit, (2) the production of nitrate is desired as an indicator of a stable and healthy biomass or as an electron carrier molecule under an anoxic condition to degrade soluble cBOD for the control undesired, filamentous organism growth, and (3) operational conditions simply favor nitrification, that is, the SBR "slips" into nitrification.

There are several major operational conditions that affect the activity and growth of nitrifying bacteria and consequently the ability of an SBR to nitrify. These conditions include:

- Mean cell residence time (MCRT)
- Temperature
- Dissolved oxygen
- pH
- Alkalinity
- Toxicity

The most important major conditions are MCRT and temperature. Additional operational conditions that influence nitrification include:

- High influent cBOD
- High influent nBOD
- React time
- Phosphorus deficiency

Nitrification may be complete or incomplete (partial). Complete nitrification occurs when the mixed liquor filtrate after the React Phase contains <1 mg/L ammonium, <1 mg/L nitrite, and as much as possible nitrate. A filtrate sample of mixed liquor is obtained by passing the mixed liquor through a Whatman® No. 4 filter paper. The liquid passing through the filter paper is the filtrate.

Because there are two groups of nitrifying bacteria and two biochemical reactions involved in nitrification, there are four forms of incomplete nitrification. Biochemical reactions are chemical reactions that occur inside living cells. The forms of incomplete nitrification in an SBR can be identified by measuring the concentration of ammonium, nitrite, and nitrate in a sample of mixed-liquor filtrate after the React Phase (Table 8.2). By determining the form of nitrification that is occurring, it is possible to identify the operational factor responsible for its occurrence (Table 8.3).

TABLE 8.2 Forms of Nitrification

Form of Nitrification	SBR Filtrate Sample after React Phase		
	Ammonium (mg/L)	Nitrite (mg/L)	Nitrate (mg/L)
Complete	<1	<1	>1
Incomplete No. 1	<1	>1	<1
Incomplete No. 2	>1	<1	>1
Incomplete No. 3	<1	>1	>1
Incomplete No. 4	>1	>1	>1

TABLE 8.3 Operational Factors Responsible for Incomplete Nitrification

Form of Incomplete Nitrification	Operational Factors Responsible for Incomplete Nitrification
Incomplete No. 1	Theoretical (not likely to occur)
Incomplete No. 2	Limiting factor[a]
Incomplete No. 3	Usually cold temperature
Incomplete No. 4	Limiting factor[a] or cold temperature

[a]Limiting factors include low dissolved oxygen, pH changes, slug discharge of soluble cBOD, and inhibition/toxicity.

Some forms of incomplete nitrification (No. 3 and No. 4) result in the production and accumulation of nitrite. The accumulation of nitrite is problematic. Nitrite is toxic and exerts a significant chlorine demand that interferes with the control of undesired filamentous organism growth via chlorination and disinfection of the effluent via chlorination. This demand for chlorine is known as the "chlorine sponge," "nitrite kick," and "nitrite lock."

MCRT AND TEMPERATURE

Because nitrifying bacteria are temperature-sensitive, nitrification also is temperature-sensitive. There is a significant reduction in the rate of nitrification with decreasing wastewater temperature and, conversely, a significant acceleration in the rate of nitrification with increasing wastewater temperature. Therefore, a smaller population of nitrifying bacteria is needed to achieve acceptable nitrification with increasing wastewater temperature, while a larger population of nitrifying bacteria is needed to achieve acceptable nitrification during cold wastewater temperature.

Because nitrifying bacteria grow much more slowly than organotrophic bacteria that remove cBOD (Table 8.4), it is possible to waste a significant number of nitrifying bacteria from the SBR at a higher rate than their growth rate. Therefore, adequate time (usually 8 days MCRT or higher) must be provided for nitrifying bacteria to increase in number and oxidize ammonium to nitrate.

Bacteria in an SBR are a component of the solids or mixed-liquor suspended solids (MLSS). Because bacteria are volatile (change quickly to vapor at 550 °C), they are referred to as the mixed-liquor volatile suspended solids (MLVSS) and represent approximately 75% of the MLSS. Nitrifying bacteria represent approximately 10% of all bacteria in an SBR. The number of nitrifying bacteria in an SBR can be increased or decreased by increasing or decreasing, respectively, the MLSS concentration.

The MLSS concentration is a function of the MCRT. With increasing MCRT, more bacteria are present in an SBR; with decreasing MCRT, fewer bacteria are present in the SBR. The relationship between temperature and MCRT is provided in Table 8.5, while significant temperature values and their impact upon nitrification are provided in Table 8.6.

With all operational conditions favorable for nitrification, the rate of nitrification is most rapid at 30 °C. However, nitrification in an SBR may improve at temperature values greater than 30 °C, if the increased temperature promotes more rapid removal of soluble cBOD. When soluble cBOD is removed more rapidly, more aeration time during React Phase is available for nitrification.

TABLE 8.4   Optimum Generation Time for Significant Groups of Wastewater Bacteria

Bacterial Group	Generation Time
Organotrophs, activated sludge	15–30 minutes
Nitrifying bacteria, activated sludge	2–3 days
Methane-forming bacteria	3–30 days

TABLE 8.5   Temperature and MCRT Recommended for Nitrification

Temperature (°C)	MCRT Recommended for Nitrification
10	30 days
15	20 days
20	15 days
25	10 days
30	7 days

TABLE 8.6   Impact of Temperature upon Nitrification

Temperature (°C)	Impact upon Nitrification
30	Maximum rate of nitrification
15	Loss of 50% of maximum rate
10	Loss of 20% of maximum rate
8	Nitrification inhibited

TABLE 8.7   Oxygen Consumption (Theoretical) Required for Nitrification

Biochemical Reaction	Dissolved Oxygen Consumed (pounds)
1 pound NH_4^+ to 1 pound NO_2^-	3.43
1 pound NO_2^- to 1 pound NO_3^-	1.14
1 pound NH_4^+ to 1 pound NO_3^-	4.57

MLVSS concentrations greater than 2000 mg/L are commonly used to nitrify, and higher MLVSS concentrations often are required with cold wastewater temperatures (≤15 °C). With increasing MLVSS concentration the food-to-microorganism (F/M) ratio of an SBR decreases. An F/M of ≤0.08 promotes nitrification.

DISSOLVED OXYGEN

Nitrifying bacteria are strict aerobes, and nitrification is a strict aerobic reaction. Nitrification consumes a large quantity of dissolved oxygen. Approximately 4.6 pounds of oxygen are consumed for each pound of ammonium that is oxidized to nitrate (Table 8.7).

Nitrification is most rapid between a dissolved oxygen concentration of 2.0 mg/L and 3.0 mg/L. The maximum rate of nitrification is considered to occur at approximately 3.0 mg/L. Nitrification in an SBR may improve at dissolved oxygen values greater than 3 mg/L, if the increased dissolved oxygen concentration promotes more rapid removal of soluble cBOD. When soluble cBOD is removed more rapidly, more aeration time is available for nitrification. Significant dissolved oxygen values or ranges in dissolved oxygen values that impact nitrification are provided in Table 8.8.

TABLE 8.8 Dissolved Oxygen Concentration and Nitrification

Dissolved Oxygen Concentration (mg/L)	Impact upon Nitrification
<0.5	Little, if any, nitrification occurs
0.5–0.9	Nitrification occurs, usually inefficient
2.0–2.9	Significant nitrification occurs
3.0	Maximum rate of nitrification occurs
>3.0	Improvement in nitrification occurs, if soluble cBOD is removed more rapidly

pH AND ALKALINITY

Most SBRs nitrify successfully at a near-neutral pH, 6.8–7.2. Some SBRs operate at an increased pH, 7.6–7.8, to promote more favorable conditions for nitrification. However, increases in pH values above 7.8 are not recommended for the following reasons:

- Undesired growth of *Microthrix parvicella*
- Loss of floc formation and organotrophic enzymatic activity
- Reduction in escape of carbon dioxide from the mixed liquor

Alkalinity is destroyed in an SBR during nitrification, and consequently the pH drops. Approximately 7.14 mg of alkalinity as calcium carbonate ($CaCO_3$) is destroyed per milligram of ammonium oxidized. Alkalinity is lost in an SBR, because (1) it is used as a carbon source by nitrifying bacteria to produce new bacterial cells (sludge production) and (2) it is destroyed by the production of nitrous acid (HNO_2) in the oxidation of ammonium to nitrite [Eq. (8.3)]. Some of the nitrite produced from the oxidation of ammonium combines with hydrogen protons (H^+) also produced from the oxidation of ammonium to form free nitrous acid [Eq. (8.4)]. Free nitrous acid destroys alkalinity and is toxicity to nitrifying bacteria.

$$NH_4^+ + 1.5O_2 \xrightarrow{\text{AOB}} 2H^+ + NO_2^- + 2H_2O \tag{8.3}$$

$$H^+ + NO_2^- \rightarrow HNO_2 \tag{8.4}$$

The loss of alkalinity is responsible for two significant operational problems. First, without adequate alkalinity, nitrifying bacteria can no longer oxidize ammonium or nitrite, that is, nitrification stops. Second, with the loss of alkalinity the pH in the SBR decreases and the decrease in pH inhibits nitrification. Therefore, it is necessary to ensure an adequate amount of alkalinity in an SBR at all times.

Adequate alkalinity is considered to be ≥50 mg/L after complete nitrification. There are several chemical compounds or alkalis that are suitable for adding alkalinity for nitrification. These alkalis are listed in Table 8.9.

TOXICITY

There are several forms of inhibition or toxicity that affect nitrifying bacteria. Generally, whatever inhibits the organotrophic (cBOD-removing) bacteria also

TABLE 8.9 Alkalis That Are Suitable for Alkalinity Addition

Alkali	Common Name	Formula	$CaCO_3$ Equivalent
Calcium carbonate	Calcite, limestone	$CaCO_3$	1.00
Calcium hydroxide	Lime	$Ca(OH)_2$	1.35
Sodium bicarbonate	Baking soda	$NaHCO_3$	0.60
Sodium carbonate	Soda ash	Na_2CO_3	0.94
Sodium hydroxide	Caustic soda	$NaOH$	1.25

inhibits the nitrifying bacteria but at a much lower concentration. However, there are two unique forms of inhibition that occur for nitrifying bacteria. These forms are (1) soluble cBOD inhibition and (2) substrate toxicity.

Soluble cBOD inhibition occurs because some simplistic forms of cBOD such as methanol (CH_3OH), ethanol (CH_3CH_2OH), propanol ($CH_3CH_2CH_2OH$), and butanol ($CH_3CH_2CH_2CH_2OH$) have not been adequately degraded by the organotrophic bacteria. These forms of soluble cBOD enter nitrifying bacteria and complex (tie up) their enzymatic machinery and prevent nitrification. Therefore, to prevent soluble cBOD inhibition, the cBOD in an SBR should be <40 mg/L to ensure the occurrence of nitrification.

Substrate toxicity occurs because an energy substrate for nitrifying bacteria, ammonium or nitrite, accumulates to an undesired level in an SBR. When ammonium is present in an SBR at concentrations ≥480 mg/L, one of the following two conditions may occur that results in the inhibition of nitrification:

- The ammonium accumulates, and the pH of the SBR increases. This increase in pH results in the production of free ammonia (NH_3). Free ammonia is toxic.
- The ammonium is quickly oxidized to nitrite, and free nitrous acid (HNO_2) is produced in a large quantity. The production of nitrous acid results in a decrease in pH that accelerates the production of more nitrous acid. Free nitrous acid is toxic.

Substrate toxicity can be prevented by three operational measures. First, the slug discharge of ammonium or nitrogen-containing wastes to an SBR should be prevented. Second, nBOD loading should be equalized. Third, the pH of an SBR should be maintained at near neutral pH (6.8–7.2).

RETENTION TIME (REACT PHASE)

With increasing retention time under aeration (React Phase), nitrification is promoted. The increase in aeration time provides for more time for nitrification after the reduction in soluble cBOD. Nitrifying bacteria are strict aerobes, and the biochemical reactions involved in nitrification are aerobic.

PHOSPHORUS DEFICIENCY

Phosphorus is an essential nutrient that is required for the activity and growth of all bacteria. However, nitrifying bacteria must compete with organotrophic bacteria for phosphorus. When a marginal or deficient condition exists for reactive phosphorus or orthophosphate ($H_2PO_4^-/HPO_4^{2-}$), nitrifying bacteria cannot nitrify. Therefore, a residual value of ≥0.5 mg/L of reactive phosphorus should be present in the mixed-liquor filtrate immediately after the React Phase to ensure that an adequate concentration of phosphorus is available for nitrifying bacteria. This concern for a phosphorus deficiency should be addressed when chemical phosphorus removal is practiced.

TROUBLESHOOTING KEY FOR THE IDENTIFICATION OF OPERATIONAL FACTORS RESPONSIBLE FOR THE LOSS OF NITRIFICATION

1. Is the MLVSS low?

   Yes   See 2
   No   See 3

2. Increase MLVSS to concentration necessary for warm or cold wastewater temperature operation of the SBR. MCRT or F/M may be used as a guideline.

3. Is the dissolved oxygen concentration low during React Phase?

   Yes   See 4
   No   See 5

4. Increase the dissolved oxygen concentration to 3 mg/L. If it is difficult to increase the dissolved oxygen concentration in the SBR, identify the cause for the low dissolved oxygen concentration.

5. Is the pH operating range of the SBR near neutral (6.8–7.2) or has the pH dropped significantly or changed more than 0.3 standard units over the last 24 hours?

   Yes   See 6
   No   See 7

6. Adjust and stabilize the pH. A pH drop may be due to the loss of alkalinity through nitrification. Swings in pH may be due to changes in influent composition or chemicals added to the SBR. Identify the source of the pH change and neutralize or adjust the influent pH.

7. Has inhibition or toxicity occurred?

   Yes   See 8
   No   See 9

8. If inhibition is due to the presence of cBOD, extend Aerated Fill Phase or React Phase or increase MLVSS. If substrate toxicity occurs, maintain a near-neutral pH (6.8–7.2) in the SBR and identify and regulate the discharge source of

ammonium or nitrogen-containing compounds. If toxicity is due to other organic or inorganic wastes, identify and regulate the discharge source and increase MLVSS inventory.

9. Is there inadequate aeration time for cBOD degradation and nitrification?

Yes   See 10
No   See 11

10. Increase aeration time.

11. Determine if a phosphorus deficiency exists and add a phosphorus-containing compound such as phosphoric acid to the SBR until a residual orthophosphate ≥0.5 mg/L is obtained in the mixed-liquor filtrate after React Phase.

Check List for the Identification of Operational Factors Responsible for Loss of Nitrification

Operational Factor	√ If Monitored	√ If Possible Factor
Low MLVSS		
Low dissolved oxygen		
Low pH		
Low alkalinity		
Inhibition/toxicity		
Excess soluble cBOD		
Insufficient aeration time		
Phosphorus deficiency		

9

Troubleshooting Denitrification

BACKGROUND

Denitrification is typically the use of nitrate (NO_3^-) or atypically nitrite (NO_2^-) by facultative anaerobic (denitrifying) bacteria to degrade soluble cBOD. Denitrifying bacteria prefer the use of molecular oxygen (O_2) to degrade soluble cBOD but have the enzymatic ability to use nitrate in the absence of molecular oxygen or in the presence of an oxygen gradient (Figure 9.1).

Nitrate is produced in an SBR during nitrification. However, nitrate or nitrite may enter an SBR from an industrial discharge (Table 9.1) or through the use of a malodor control compound. Compounds commonly used to control malodor production include sodium nitrate ($NaNO_3$) and calcium nitrate ($Ca(NO_3)_2$).

When denitrification occurs, soluble cBOD is degraded and the nitrogen in nitrate is released to the atmosphere in the form of molecular nitrogen (N_2) and not discharged in the SBR decant [Eq. (9.1)]. Denitrification occurs in an SBR during (1) the Mix Fill Phase and/or (2) after the React Phase with carbon addition. If denitrification occurs after the React Phase, a short aeration or mixing period is required following denitrification in order to stripped entrapped gases from the floating solids. This stripping action permits the solids to settle in the SBR.

$$6NO_3^- + 5CHOH \rightarrow 3N + 5CO_2 + 7H_2O + 6OH^- \qquad (9.1)$$

Commonly used chemicals for carbon addition include acetate (CH_3COOH), sodium acetate ($NaOOCCH_3$), ethanol (CH_3CH_2OH), glucose ($C_6H_{12}O_6$), and methanol (CH_3OH). Domestic wastewater contains soluble cBOD that is used as a carbon source during Mix Fill Phase.

Troubleshooting the Sequencing Batch Reactor, by Michael H. Gerardi
Copyright © 2010 by John Wiley & Sons, Inc.

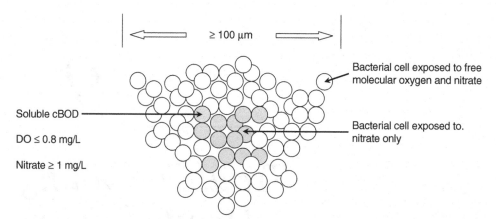

Figure 9.1 *Oxygen gradient. An oxygen gradient is established across a floc particle, if the floc particle is >100 μm in size and the dissolved oxygen outside the floc particle is ≤0.8 mg/L. Under this condition a measurable dissolved oxygen concentration is present, but the dissolved oxygen cannot penetrate to the core of the floc particle. Bacteria on the perimeter of the floc particle experience an oxic (aerobic) environment and degrade cBOD using dissolved oxygen, while bacteria in the core of the floc particle experience an anoxic condition in the presence of nitrate and degrade cBOD using nitrate. Nitrate is not used in the perimeter of the floc particle due to the preference for dissolved oxygen over nitrate by the perimeter bacteria.*

TABLE 9.1   Industrial Wastewaters that Contain Nitrate or Nitrite

Industrial Wastewater	Nitrate	Nitrite
Corrosion inhibitor		X
Leachate, pretreated	X	X
Meat, flavoring	X	
Meat, pretreated	X	X
Steel	X	X

There are several operational conditions that affect the growth and activity of denitrifying bacteria, their ability to use nitrate, and consequently denitrification in the SBR. These conditions include:

- Abundance and activity of denitrifying bacteria
- Temperature
- Nutrients
- Oxidation–reduction potential
- Denitrification time
- Soluble cBOD

The most important operational condition that affects denitrification is the quantity of available soluble cBOD or carbon.

ABUNDANCE AND ACTIVITY OF DENITRIFYING BACTERIA

Denitrifying bacteria utilize nitrate in the absence of oxygen or the presence of an oxygen gradient to degrade soluble cBOD. Denitrifying bacteria are present in millions per milliliter of mixed liquor bulk solution and billions per gram of floc particles (solids). Their generation time in the mixed liquor is approximately 15–30 minutes.

Denitrifying bacteria enter an SBR through fecal waste and inflow and infiltration (I/I) as soil and water organisms. An adequate number of denitrifying bacteria are present for denitrification when the MLVSS are >1000 mg/L. Denitrifying bacteria are more tolerant of adverse operational conditions than nitrifying bacteria. Therefore, nitrification becomes sluggish or terminates before denitrification becomes sluggish. Commonly reported groups or genera of denitrifying bacteria include *Alcaligenes*, *Achromobacter*, *Bacillus*, and *Pseudomonas*.

TEMPERATURE

The activity of denitrifying bacteria increases with increasing wastewater temperature, and it decreases with decreasing wastewater temperature. Denitrification is more easily achieved with increasing wastewater temperature for two reasons. First, with increasing bacterial activity, dissolved oxygen is more rapidly consumed, allowing for more rapid use of nitrate. Second, with increasing wastewater temperature, less oxygen dissolves in the wastewater. This provides for a lower dissolved oxygen concentration in the SBR after React Phase.

NUTRIENTS

Because the anoxic degradation of cBOD—that is, the use of nitrate to degrade soluble cBOD—results in decreased bacterial growth or sludge production, smaller quantities of nitrogen and phosphorus are needed as compared to aerobic degradation of cBOD [Eqs. (9.2) and (9.3)]. Therefore, the residual values for nitrogen ≥ 1.0 mg/L NH_4^+-N or ≥ 3 mg/L NO_3^--N and ≥ 0.5 mg/L HPO_4^{2-}-P) and phosphorus that are adequate for aerobic degradation of cBOD after React Phase are more than adequate for anoxic degradation of cBOD.

$$1 \text{ lb } C_6H_{12}O_6 + O_2 \xrightarrow{\text{facultative anaerobic bacteria}} 0.6 \text{ lb sludge} + H_2O + CO_2 \tag{9.2}$$

$$1 \text{ lb } C_6H_{12}O_6 + NO_3^- \xrightarrow{\text{facultative anaerobic bacteria}} \\ 0.4 \text{ lb sludge} + H_2O + CO_2 + N_2 \tag{9.3}$$

When nitrate is used to degrade cBOD, less carbon from the cBOD goes into cell growth (sludge production) and more carbon from the cBOD goes into carbon dioxide (CO_2) as compared to the use of free molecular oxygen. Therefore, not all of the carbon dioxide produced during denitrification dissolves in the wastewater.

TABLE 9.2   Guideline Redox Values for Biochemical
Reactions

Biochemical Reaction	Redox Range (mV)
Nitrification	+100 to +350
cBOD degradation with O_2	+50 to +250
Biological phosphorus uptake	+25 to +250
Denitrification	+50 to −50
Sulfide formation	−50 to −250
Biological phosphorus release	−100 to −250
Acid formation (fermentation)	−100 to −225
Methane production	−175 to −400

Some escapes to the atmosphere and, unfortunately, at times becomes entrapped in solids that float to the surface of the tank.

OXIDATION–REDUCTION POTENTIAL

Biologically, oxidation–reduction potential or redox (ORP) is a measurement of the capacity of an aqueous environment to conduct specific biochemical reactions—for example, the use of molecular oxygen or nitrate to degrade soluble cBOD (Table 9.2). Oxidation–reduction potential is measured with an ORP probe and recorded as millivolts (mV). The millivolts may be positive or negative. Denitrification or the use of nitrate to degrade cBOD generally occurs within the ORP range of +50 to −50 mV.

DISSOLVED OXYGEN

Because denitrifying bacteria produce more offspring (new bacterial cells or sludge) using molecular oxygen to degrade substrate (soluble cBOD) than using nitrate, the presence of molecular oxygen does not permit denitrification. Therefore, dissolved oxygen must not be present or an oxygen gradient must be present in order for denitrification to occur. An oxygen gradient exists when dissolved oxygen cannot penetrate to the core of the solids or floc particles. An oxygen gradient exists when the dissolved oxygen outside the floc particles is ≤0.8 mg/L and the size of the floc particles is >100 μm. Because most activated sludge processes operate at a high MCRT to develop an adequate population of nitrifying bacteria for nitrification, these plants grow a relatively large number of filamentous organisms. These organisms provide for the development of numerous medium (150–500 μm) and large (>500 μm) floc particles and an oxygen gradient if the dissolved oxygen value in the bulk solution is ≤0.8 mg/L. Therefore, in the presence of a residual quantity of oxygen, denitrification can occur.

DENITRIFICATION TIME

Denitrification time may occur during a Mix Fill Phase or at the end of the React Phase or Settle Phase. Denitrification time should be adjusted to provide for adequate removal of nitrate, reduction in dissolved oxygen concentration, and control of undesired filamentous organism growth.

SOLUBLE CBOD

The most important operational condition affecting denitrification is soluble cBOD. The greater the amount of soluble cBOD that is present, the more rapidly residual dissolved oxygen is used by aerobic and facultative anaerobic bacteria to degrade soluble cBOD and the more rapidly nitrate is used by the facultative anaerobic bacteria to degrade soluble cBOD. Approximately three parts of soluble cBOD (carbon) are required for each part of nitrate to be removed by facultative anaerobic bacteria. Commonly used chemicals for carbon addition include acetate (CH_3COOH), sodium acetate ($NaOOCCH_3$), ethanol (CH_3CH_2OH), glucose ($C_6H_{12}O_6$), and methanol (CH_3OH).

Several significant benefits of denitrification include a higher quality of treated wastewater (less nitrogen in the decant) and recovery of approximately 50% of the alkalinity lost during nitrification. Another significant benefit is recovery of approximately 33% of the energy spent for aeration, if aeration is terminated and nitrates are used to degrade cBOD.

Denitrifying bacteria also compete for fatty acids that are produced during biological phosphorus release under an anaerobic/fermentative condition (Static Fill Phase). This competition is due to the presence of two groups of denitrifying bacteria. These groups consist of (1) cBOD-removing bacteria that have a primary affinity for easily degradable cBOD, especially fatty acids (acetate, propionate and butyrate) that are produced during biological phosphorus release, and (2) cBOD-removing bacteria that are "phosphorus-removing" bacteria.

Therefore, if nitrates are present at the beginning of the Static Fill Phase, the first group of denitrifying bacteria consume fatty acids but do not store the fatty acids for subsequent biological phosphorus uptake. This results in a reduction in the ability of Poly-P bacteria (phosphorus-accumulating organisms) to remove phosphorus from the bulk solution. Denitrifying bacteria in the second group are phosphorus-removing bacteria, but their activity is difficult to predict. Therefore, well-controlled denitrification is recommended, if biological phosphorus removal is practiced in an SBR.

TROUBLESHOOTING KEY FOR THE IDENTIFICATION OF OPERATIONAL FACTORS RESPONSIBLE FOR LOSS OF DENITRIFICATION

1. Is the MLVSS low?

    Yes   See 2
    No   See 3

2. Increase MLVSS as needed for cold or warm wastewater temperature.

3. Does a nutrient deficiency exist for phosphorus in the mixed liquor?

    Yes   See 4
    No   See 5

4. Add a phosphorus-containing compound such as phosphoric acid to the SBR until a residual orthophosphate-phosphorus value of ≥0.5 mg/L is maintained in the SBR filtrate after React Phase.

5. Is the ORP value acceptable for denitrification?

Yes   See 6
No   See 7

6. ORP values ≥+50 mV typically are indicative of the presence of excess dissolved oxygen that does not permit denitrification. The dissolved oxygen must be reduced or removed either by adjusting the aeration time or dissolved oxygen concentration during aeration. Adding additional soluble cBOD (carbon) during the denitrification period helps to reduce dissolved oxygen concentration and accelerate denitrification.

7. Is there residual dissolved oxygen?

Yes   See 8
No   See 9

8. Dissolved oxygen must be reduced or removed in order for denitrification to occur. Dissolved oxygen may be reduced or removed by adjusting the aeration time or dissolved oxygen concentration during the React Phase or adding addition soluble cBOD (carbon) during the denitrification period.

9. An insufficient quantity of soluble cBOD was not available for adequate denitrification. If the cBOD remaining in the SBR after React Phase is not adequate for desired denitrification, then soluble cBOD must be added to the SBR when denitrifying. If adequate cBOD is present, but denitrification is not sufficient to comply with a permit discharge limit for total nitrogen, then more time must be provided for ammonification and nitrification, and the denitrification period may need to be extended or two denitrification periods may be used.

Check List for the Identification of Operational Factors Responsible for Loss of Denitrification

Operational Factor	√ If Monitored	√ If Possible Factor
Low MLVSS		
Phosphorus deficiency[a]		
Inappropriate ORP value		
Dissolved oxygen too high		
Lack of adequate soluble cBOD		
Inadequate denitrification time		

[a]To determine the presence of an adequate quantity of phosphorus in the SBR, a grab sample of mixed liquor should be taken immediately after the denitrification (anoxic) period. The sample should be filtered through a Whatman® No. 4 filter paper. The filtrate should be tested for orthophosphate-phosphorus. If the concentration of orthophosphate-phosphorus is ≥0.5 mg/L, a nutrient deficiency for phosphorus does not exist. However, in the presence of toxicity bacterial cells do not degrade soluble cBOD and do not use orthophosphate. Therefore, an acceptable specific oxygen uptake rate (SOUR) should be obtained to exclude toxicity as a possible condition permitting the presence of an acceptable residual of orthophosphate-phosphorus in the filtrate.

10

Troubleshooting High Decant BOD

BACKGROUND

Biochemical oxygen demand (BOD) consists of carbonaceous and nitrogenous wastes that can be degraded by microorganisms, primarily bacteria, with the use of dissolved oxygen (DO). BOD represents the quantity (mg/L) of dissolved oxygen that is consumed by microorganisms to degrade (oxidize) carbonaceous and nitrogenous wastes. The BOD is used to determine the strength of the wastewater, that is, the higher the BOD value (mg/L), the greater the quantity of dissolved oxygen (mg/L) that is demanded or consumed by the microorganisms to degrade the wastes and the greater the pollution. The BOD of the wastewater typically is performed over a 5-day period, and the 5-day period is commonly referenced as a subscript "5" in BOD_5.

The oxidation of carbonaceous and nitrogenous wastes occurs through chemical reactions inside the microorganisms. Chemical reactions inside microorganisms are known as biochemical reactions. Therefore, the quantity of oxygen demanded by microorganisms to oxidize carbonaceous and nitrogenous compounds in wastewater is known as the biochemical oxygen demand.

Carbonaceous or organic wastes consist of compounds that contain carbon and hydrogen. Carbonaceous wastes make up the carbonaceous BOD or cBOD. Examples of carbonaceous compounds include acids such as acetate or vinegar (CH_3COOH), alcohols such as ethanol (CH_3CH_2OH), and sugars such as glucose ($C_6H_{12}O_6$). Nitrogenous wastes consist of nitrogen-containing compounds that can be degraded or oxidized by nitrifying bacteria. There are only two nitrogen-containing wastes that are oxidized by nitrifying bacteria. These wastes consist of ammonium (NH_4^+) and nitrite (NO_2^-).

Troubleshooting the Sequencing Batch Reactor, by Michael H. Gerardi
Copyright © 2010 by John Wiley & Sons, Inc.

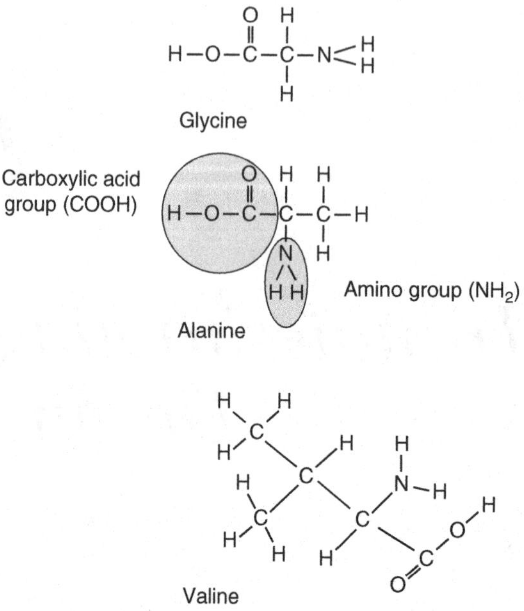

Figure 10.1 *Examples of amino acids. There are 20 naturally occurring amino acids. All amino acids contain a carboxylic acid group (-COOH) and an amino group (-NH₂). Examples of several simple aliphatic amino acids include glycine (H₂NCH₂COOH), alanine (CH₃CH(NH₂)COOH), and valine (HOOCCHNH₂C(CH₃)₂).*

TABLE 10.1 Types of BOD

Type of BOD	Relative Rate of Degradation
Total BOD	Based upon composition of BOD
Carbonaceous BOD	Based upon composition of cBOD
Particulate BOD	Slow rate of degradation
Colloidal BOD	Slow rate of degradation
Soluble carbonaceous BOD	Rapid rate of degradation
Nitrogenous BOD	Rapid rate of degradation after significant soluble cBOD degradation

There is a group of carbonaceous BOD that contributes to nBOD. This group consists of organic-nitrogen compounds such as amino acids (Figure 10.1) and proteins. Amino acids are the building blocks for proteins. Although these forms contain carbon and hydrogen and are properly referred to as cBOD compounds, they also contain nitrogen in the form of amino groups (-NH₂). When organic-nitrogen compounds are degraded, the amino groups are released. The release of amino groups is known as ammonification and results in the production of ammonium at pH values of <9.4.

There are several types of cBOD (Table 10.1) and only one type of nBOD. Carbonaceous BOD consists of particulate, colloidal, and soluble types. Particulate and colloidal cBOD degrade slowly, while soluble cBOD degrades rapidly. An example of particulate BOD is cellulose. Cellulose (Figure 10.2) is an insoluble starch and can be easily observed as the "stringy" material in celery or the translu-

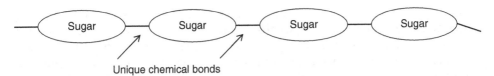

Figure 10.2 *Cellulose. Cellulose is an insoluble starch. Cellulose consists of a chain of simple sugars (glucose) that are bonded together by unique chemical bonds. Only those bacteria that produce the enzyme that breaks these chemical bonds are able to hydrolyze cellulose into soluble simple sugars that can be absorbed by numerous bacteria and degraded.*

cent coating over corn kernels and peas. Colloidal cBOD consists of complex molecules such as proteins that have a large surface area, are insoluble in wastewater, and are suspended in wastewater. Soluble cBOD consists of simple molecules such as acids, alcohols, and sugars.

Regulatory agencies require that wastewater treatment facilities reduce the quantity of BOD or cBOD in the wastewater before it is discharged. Unfortunately, undesired changes in the quantity and/or quality (composition) of the influent or undesired conditions within an SBR can contribute to excess BOD or cBOD in the decent. There are several operational conditions that contribute to excess BOD. These conditions include:

- Biodegradable effluent (floatable and suspended) solids
- Change in
    - Aeration time
    - F/M
    - Hydraulic loading
    - MLVSS
    - Organic loading
    - pH
- Total dissolved solids (TDS)
- Depressed temperature
- Loss of nitrification
- Low dissolved oxygen level
- Nutrient deficiency
- Over-aeration
- Toxicity
- High effluent TSS

BIODEGRADABLE EFFLUENT (FLOATABLE AND SUSPENDED) SOLIDS

Solids that do not settle during the Settle Phase may become effluent solids. If the effluent solids are biodegradable—that is, contribute to BOD—then an increase in decant BOD occurs. Effluent solids consist of floatable solids and suspended solids. Floatable solids include foam and scum, while suspended solids consist of dispersed cells, floc particles, filamentous organisms, and particulate material. Floatable and suspended solids usually contain adsorbed colloids.

Floatable solids consisting of foam and scum are reviewed in Chapter 13, which is entitled "Troubleshooting Foam and Scum Production." Scum represents the die-off of a relatively large number of bacteria due to seasonal changes in wastewater temperature, toxicity, and starvation. Foam and scum are biodegradable and contribute to BOD.

CHANGE IN AERATION TIME

A change or decrease in React Phase may contribute to an increase in effluent BOD. A decrease in aeration time may not provide for adequate degradation of particulate and colloidal material and may contribute to a loss of adequate nitrification. Nitrification, the oxidation of ammonium to nitrate, reduces nBOD, which is a component of BOD.

The wastewater should have a dissolved oxygen value of zero or near zero mg/L during most of the Static (anaerobic/fermentative) Fill Phase and should have a dissolved oxygen value of ≤0.8 mg/L during most of the Mix (anoxic) Fill Phase. The anaerobic and/or anoxic portions of the Fill Phase are for short periods of time. The dissolved oxygen during the React Phase should gradually increase to 3–4 mg/L by the end of the React Phase. Dissolved oxygen of >4 mg/L at the end of the React Phase may be considered an excess quantity. If so, either aeration time or the aeration rate should be reduced.

CHANGE IN F/M

The F/M of an SBR is the ratio of food (pounds of BOD) that enters the SBR to the number of microorganisms (pounds of mixed liquor volatile suspended solids) that are present in the SBR. Treatment systems typically operate at specific values or range of values for F/M (for example, ≤0.08 for nitrification), and changes in F/M can affect treatment efficiency.

The F/M increases or decreases with changes in BOD and MLVSS concentration. If an excess quantity of BOD enters the SBR (for example, a slug discharge of soluble BOD) and it is not adequately degraded, it "breaks through" the SBR in the decant. If a deficient quantity of BOD enters the SBR, many bacteria may die due to starvation. The death of the bacteria results in autolysis ("self-splitting") and loss of cellular components (BOD) in the decant. These components also contribute to scum formation.

CHANGE IN HYDRAULIC LOADING

A significant change in hydraulic loading may result in an increase in BOD in the decant in a continuous flow SBR or single-feed SBR. An increase in hydraulic loading may occur due to a mechanical failure of the treatment system to equalize the flow, an increase in the quantity of wastewater discharged to the SBR, or the occurrence of inflow and infiltration (I/I) that requires a reduction in phase times (shorter cycle time), especially React Phase. An increase in hydraulic loading that decreases the time that the BOD is under aeration results in excess BOD in the decant.

CHANGE IN ORGANIC LOADING

A change in organic loading may occur as an increase in the quantity of BOD received or the presence of more soluble BOD. For example, a slug discharge of BOD, especially soluble cBOD, may result in an excess quantity of BOD in the decant, if React Phase is not adjusted to treat the excess BOD.

CHANGE IN pH

Increases and decreases in pH above or below (respectively), the optimum pH operational range (6.8–7.2) may contribute to operational problems including loss of BOD treatment efficiency. A rapid change in pH above or below optimum pH operating range adversely affects enzymatic activity of the bacteria and their ability to oxidize BOD, especially nBOD. The BOD that is not oxidized in the SBR leaves the SBR in the decant.

A change in pH also adversely affects floc formation. At undesired pH values, floc particles become weak and buoyant. These particles are easily sheared and float, resulting in the loss of solids (colloidal and particulate BOD) in the decant. The solids from the floc particles also represent a loss of bacteria that oxidize BOD.

CHANGE IN TOTAL DISSOLVED SOLIDS

Increases in total dissolved solids (TDS) adversely affect the structure of the bacterial cell and its ability to treat BOD and form dense floc particles. TDS concentrations >5000 mg/L contribute to excess BOD in the decant as a result of damage to cellular structures. Although bacterial cells can slowly acclimate to TDS concentrations >5000 mg/L, they experience much difficulty acclimating to decreases in TDS from concentrations >5000 mg/L.

An example of a rapid and potentially adverse increase in TDS is the discharge of frac water to an activated sludge process. Frac water is the water removed from natural gas drilling operations. The water contains relative high concentrations of dissolved solids, especially chloride ions and sodium ions. Fortunately, treatment processes that receive domestic and municipal wastewater seldom experience TDS problems.

DEPRESSED TEMPERATURE

With decreasing wastewater temperature bacterial activity in an SBR decreases. Therefore, it is necessary to increase the MLVSS concentration during depressed temperatures to compensate for decreasing activity. At 14 °C approximately 50% of the ability of the SBR to oxidize cBOD is lost, while approximately 50% of the ability of the SBR to oxidize nBOD is lost at 15 °C.

LOSS OF NITRIFICATION

Nitrification is the oxidation of ammonium to nitrate. Nitrification reduces the quantity of nBOD in the wastewater and the decant. By reducing the quantity of nBOD, total BOD also is reduced in the wastewater and decant. If loss of nitrification occurs, then an increase in BOD occurs. Operational conditions responsible for the loss of nitrification are reviewed in Chapter 8, entitled "Troubleshooting Nitrification."

LOW DISSOLVED OXYGEN LEVEL

A low dissolved oxygen level permits undesired operational conditions in an SBR. These conditions include (1) the undesired growth of filamentous organisms including *Haliscomenobacter hydrossis*, *Microthrix parvicella*, *Sphaerotilus natans*, and type 1701, (2) unacceptable endogenous respiration or inefficient degradation of stored food, (3) production of weak and buoyant floc particles, (4) loss of nitrification, and (5) decreased enzymatic activity.

Undesired growth of filamentous organisms and weak and buoyant floc particles contribute to loss of fine solids (BOD) and loss of bacteria and treatment efficiency. Loss of nitrification contributes to increased nBOD in the decant, and decrease enzymatic activity contributes to increased cBOD in the decant.

NUTRIENT DEFICIENCY

With inadequate quantities of nitrogen or phosphorus, bacteria in the SBR cannot produce new cells (sludge) through the degradation of BOD. If cellular reproduction does not occur, then BOD is not oxidized. BOD that is not oxidized in the SBR may be stored in floc particles as insoluble starch granules that may be lost with the occurrence of sludge bulking. Operational conditions that are responsible for a nutrient deficiency for nitrogen or phosphorus are reviewed in Chapter 13, "Troubleshooting Foam and Scum Production."

OVER-AERATION

Over-aeration of an SBR contributes to several operational problems. These problems include (1) increased energy costs, (2) undesired nitrification and denitrification, if not required by a regulatory agency, (3) shearing action, and (4) enhanced foam production. Undesired rates of aeration also contribute to shearing action and loss of fine solids. The loss of solids represents an increase in decant BOD and a decrease in treatment efficiency. Over-aeration increases foam production. Foam lost from an SBR represents an increase in decant BOD.

TOXICITY

Toxicity may be acute or chronic. Acute toxicity occurs rapidly and lasts approximately 2–3 days. Chronic toxicity occurs slowly and may last for several weeks. The presence of toxicity may be detected by the following operational conditions:

- Loss of nitrification
- Discharge of atypical and elevated concentrations of ammonium, nitrite and orthophosphate in the decant
- Discharge of atypical and elevated concentration of total suspended solids in the decant
- Increase in decant conductivity
- Decrease in specific oxygen uptake rate (SOUR)

The specific oxygen uptake rate serves as an indicator of the activity of the biomass in the SBR. The SOUR includes the oxygen uptake rate (OUR) or respiration rate (RR) during aeration and the relative number of bacteria (MLVSS) that use the oxygen. The OUR measures the quantity of dissolved oxygen (mg/L) consumed per hour. The specific oxygen uptake rate is the OUR divided by the MLVSS.

In order to use SOUR to determine possible toxicity in an SBR, the SOUR must be performed under similar loading conditions each day—that is, the same React Phase and phase time each day. A significant decrease in SOUR may indicate toxicity. This value along with the operational conditions listed above is helpful in identifying toxicity. To perform the SOUR, the following procedure is recommended:

- Collect at least 1 L of fresh mixed liquor during the React Phase.
- Immediately after collecting the mixed liquor, stir the mixed liquor and transfer approximately 750 mL of mixed liquor to a 1-L bottle.
- Cap the 1-L bottle and thoroughly shake or aerate the mixed liquor to bring the dissolved oxygen concentration in the bottle above 5 mg/L.
- Pour a well-mixed portion of the aerated mixed liquor into a BOD bottle and fill to overflowing. If any air bubbles are trapped in the BOD bottle, tap the bottle to allow the bubbles to escape.
- If the BOD probe for the dissolved oxygen testing has a stirrer, then place the probe in the BOD bottle and note the dissolved oxygen concentration on the meter. If the dissolved oxygen probe does not have a stirrer, then place a magnetic stirring bar in the BOD bottle and place the bottle on a stirring plate. Begin stirring and note the dissolved oxygen concentration on the meter.
- Allow 30–60 seconds for the dissolved oxygen meter to stabilize.
- After the dissolved oxygen meter has stabilized, begin recording the dissolved oxygen concentration of the mixed liquor in 30-second intervals. Record

dissolved oxygen concentrations for 5–10 minutes, but do not record dissolved oxygen concentrations of <1 mg/L.

- Graph the results by plotting dissolved oxygen concentration (mg/L) on the vertical axis and time (minutes) on the horizontal axis.
- Draw a straight line that connects the majority of the points. Extend the line so it crosses the vertical and horizontal axes.
- Determine the slope of the line. The slope of the line is the change in rise (milligrams per liter of dissolved oxygen) over the change in the run (minutes) of the line or milligrams per liter of dissolved oxygen per minute (mg/L O$_2$/minute). Perhaps the easiest way to determine the slope of the line is to use the points at which the line crosses the axes. Divide the milligrams per liter of dissolved oxygen crossing the vertical axis by the time in minutes crossing the horizontal axis. The slope of the line is the oxygen uptake rate (OUR).
- Determine the specific oxygen uptake rate (SOUR) according to the following equation:

$$\text{SOUR (mg/g)/h} = \text{oxygen consumption rate (mg/L)/minute} \times 60 \text{ minutes/hour} \times 1000 \text{ mg/g/MLVSS mg/L}$$

SOUR values of >20 are "high" and may indicate that there are not enough solids (MLVSS) for the BOD loading. SOUR values from 12 to 20 are "normal" and usually produce a good BOD removal and a sludge that settles well in the SBR. SOUR values of <12 are "low" and may indicate that there are too many solids or that there has been a toxic occurrence.

STRATIFICATION

Over time, chemical addition to an SBR for operational control of alkalinity, pH, nutrients, phosphorus removal, solids capture, or other purposes may result in the build up or deposit of insoluble chemicals or sludge on the bottom of the reactor. These deposits may produce improper mixing action and, consequently, stratification of dissolved oxygen concentration (Figure 10.3). Stratification may contribute to decreased treatment efficiency.

Therefore, to detect and identify stratification in an SBR, the reactor should be tested annually, semiannually, or as needed by performing a dissolved oxygen profile across the surface and depth of the mixed liquor (Figure 10.4) to ensure uniform distribution of dissolved oxygen. If stratification is detected, the SBR should be drained and the deposits removed.

Flat-bottom, rectangular or square reactors more easily permit the accumulation of deposits than those reactors with a sloping bottom. Also, an SBR with a sloping bottom requires less time for cleaning.

GENERAL AERATION DESIGN GUIDELINES

The use of several small blowers rather than one large blower is preferred for aeration of an SBR. With multiple blowers per reactor, blowers can be taken off-line

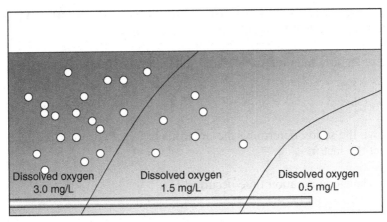

Figure 10.3 *Stratification of dissolved oxygen. Short-circuiting results in stratification of dissolved oxygen in the reactor due to the accumulation of sludge or chemicals. Stratification produces zones of high and low dissolved oxygen that are responsible for reduced treatment efficiency and undesired growth of low dissolved oxygen filamentous organisms (*Haliscomenobacter hydrossis, Microthrix parvicella, Sphaerotilus natans, *and type 1701).*

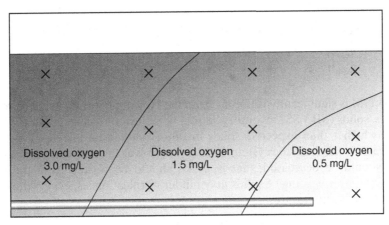

Figure 10.4 *Dissolved oxygen profile. In order to determine if stratification of dissolved oxygen has occurred in an SBR, a dissolved oxygen profile must be performed during the React Phase. The dissolved oxygen concentration of the aerated mixed liquor should be obtained at fixed distances across the surface of the SBR as well as fixed depths ("X") throughout the SBR.*

when maximum aeration is not needed or loading conditions change. If a single blower is used, it should be sized to provide maximum aeration and equipped with a variable-speed drive.

Fine-bubble membrane diffusers are preferred for aeration of an SBR. When compared to coarse-air diffusers, fine-bubble diffusers transfer more oxygen to the wastewater due to increased surface area of the fine bubbles in contact with the wastewater. Also, the greater the depth of the diffusers in the wastewater, the longer the time it takes for bubbles to reach the surface of the wastewater; that is, the greater the height of the wastewater over the diffusers, the greater the contact time and transfer time of air to the wastewater.

TROUBLESHOOTING KEY FOR THE IDENTIFICATION OF OPERATIONAL FACTORS RESPONSIBLE FOR HIGH DECANT BOD

1. Is there an accumulation of floatable solids during Decant Phase?

Yes   See 2
No   See 3

2. Remove the solids, determine the operational factor for their production, and correct the factor.

3. Has a change in aeration time occurred?

Yes   See 4
No   See 5

4. A decrease in aeration time adversely affects treatment efficiency for cBOD and nBOD removal. Determine if a decrease in aeration time was implemented and is acceptable. If not acceptable, then increase aeration time (React Phase).

5. Has a change occurred in F/M?

Yes   See 6
No   See 7

6. If the F/M is significantly different from the typical operating value or range of values, solids (MLVSS) inventory should be adjusted to correct the F/M. If the F/M is higher than expected, solids inventory should be increased, that is, do not waste solids from the system. If the F/M is lower than expected, solids inventory should be decreased, that is, solids should be wasted from the system. However, over-wasting of solids may hinder nitrification, especially during cold wastewater temperatures.

7. Has a change in quantity or quality of organic loading occurred?

Yes   See 8
No   See 9

8. Increases in BOD, especially nBOD, may not be adequately degraded with existing time for React Phase, and slowly degrading BOD also may not be adequately degraded with existing time for React Phase. Therefore, significant changes in quantity or quality of influent BOD should be identified, equalized, or minimized, and increased aeration time may be required in order to degrade the influent BOD.

9. Has the pH of the SBR changed significantly from its typical operating range?

Yes   See 10
No   See 11

10. A "safe" pH operating range for the SBR is 6.8–7.2. Changes in pH above or below this range should be adjusted with the addition of an acidic or alkali compound. The cause for the change in pH also should be identified and corrected.

11. Has depressed wastewater temperature occurred?

Yes   See 12
No   See 13

12. With decreasing wastewater temperature, bacteria become less active in removing cBOD ($\leq 14\,^\circ C$) and nBOD ($\leq 15\,^\circ C$). Therefore, in order to compensate for decreased bacterial activity, it is necessary to increase MLVSS inventory.

13. Has a loss in nBOD removal or nitrification occurred?

Yes   See 14
No   See 15

14. Nitrogenous wastes or nBOD that are not degraded in the SBR contribute to an increase in decant BOD. In order to reduce decant BOD to satisfy discharge requirements, it is necessary to restore nitrification.

15. Has a low dissolved oxygen concentration occurred?

Yes   See 16
No   See 17

16. A low dissolved oxygen concentration in the SBR reduces treatment efficiency for BOD reduction. BOD that is not degraded in the SBR contributes to an increase in decant BOD. Determine and correct the operational factor responsible for low dissolved oxygen concentration—that is, (1) mechanical problem or (2) change in concentration and/or composition of the influent.

17. Has a nutrient deficiency for nitrogen or phosphorus occurred in the SBR?

Yes   See 18
No   See 19

18. Without an adequate quantity of nitrogen or phosphorus in the SBR, bacteria cannot properly degrade BOD. Therefore, a nutrient deficiency results in an increase in decant BOD, unless the bacterial cells can successfully stored the BOD in an insoluble form in floc particles. Determine and correct the operational factor responsible for a nutrient deficiency—that is, (1) nutrient deficient industrial discharge or (2) phosphorus precipitation through chemical addition to the SBR.

19. Has over-aeration occurred?

Yes   See 20
No   See 21

20. Over-aeration of an SBR may contribute to a loss of solids through shearing action or enhance foam production. Solids lost in the Decant Phase represent BOD. Is over-aeration occurring due to (1) a mechanical problem or (2) desired increase in dissolved oxygen? If over-aeration is the result of a desired increase in dissolved oxygen, gradually reduce dissolved oxygen concentration as long as treatment performance is acceptable.

21. Toxicity in an SBR either reduces or terminates bacterial activity, including the reduction of BOD. Identify the source of toxicity and terminate the toxic discharge. Rebuild solids using influent wastewater and/or bioaugmentation products. Dead bacteria need not be wasted immediately from the SBR. The dead bacteria serve as a substrate or food source for the influent bacteria.

Check List for the Identification of Operational Factors Responsible for High Decant BOD

Operational Factor	√ If Monitored	√ If Possible Factor
Floatable solids		
Change in aeration time		
Change in F/M		
Change in organic loading		
Change in pH		
Depressed temperature		
Loss of nitrification		
Low dissolved oxygen level		
Nutrient deficiency		
Over-aeration		
Toxicity		
High TSS		

11

Troubleshooting High Decant TSS

BACKGROUND

Total suspended solids (TSS) consist of inert and living solids. Inert or nonliving solids include nonbiodegradable materials such as plastic fibers and degradable materials such as cellulose (starch). Living solids include floc bacteria, filamentous organisms, protozoa, rotifers, and free-living nematodes. Upon death, the living solids exert a dissolved oxygen demand during their decomposition. The majority of living solids that are found in an SBR decant is bacteria, and they may be present as dispersed cells (Figure 11.1) or floc particles (Figure 11.2).

In addition to suspended solids, colloids also are found in the decant and should be considered as a component of the total suspended solids. Colloids are insoluble complex molecules with a large surface area that do not settle out in wastewater unless they are adsorbed to floc particles. Some colloids such as proteins are biodegradable, while other colloids such as kaolin ("clay-like" material that is added to paper to give paper a bright white texture) are not degradable. Colloids are decanted from an SBR because they are not adsorbed to settled floc particles or they are adsorbed to floc particles that are removed during the Decant Phase.

There are several operational conditions that contribute to excess TSS in the decant (Table 11.1). These conditions include:

- Excess colloids
- Elevated temperature
- Foam production
- Increase in percent mixed-liquor volatile suspended solids (MVSS)
- Lack of ciliated protozoa

Troubleshooting the Sequencing Batch Reactor, by Michael H. Gerardi
Copyright © 2010 by John Wiley & Sons, Inc.

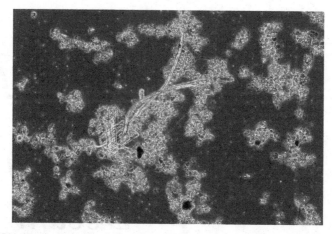

Figure 11.1   *Dispersed cells. In the bulk solution between the floc particles are many small (<10μm) spheres or floc particles. These small floc particles represent dispersed growth or dispersed cells. Their presence is usually indicative of the interruption of proper floc formation and may be associated with several operational conditions, including (1) unacceptable pH, (2) toxicity, (3) surfactants, (4) excess turbulence, and (5) low dissolved oxygen level.*

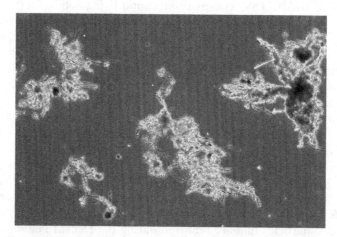

Figure 11.2   *Floc particles. In suspended growth systems such as the SBR, bacteria agglutinate as they become older. When agglutination or floc formation occurs, bacteria stick together to form floc particles. As filamentous organisms grow and extend from within the floc particles to the bulk solution, the floc bacteria grow along the lengths of the filamentous organisms and the floc particles become irregular in shape. Older bacteria that are found mostly in the core of the floc particles produce large quantities of oils that darken the core of the floc particle and give the particles their typical golden-brown color.*

TABLE 11.1   Operational Conditions Associated with the Loss of Total Suspended Solids (TSS)

Operational Condition	Description or Example
Colloids, excess	Slowly degrading proteins (slaughterhouse)
Elevated temperature	>32 °C
Foam production	Foam-producing filamentous organisms
Increase in percent MLVSS	Accumulation of fats, oils, and grease
Lack of ciliated protozoa	<100 per milliliter
Low dissolved oxygen level	<1 mg/L for 10 consecutive hours
Low pH/high pH	< 6.5/>8.5
Nutrient deficiency	Usually nitrogen or phosphorus
Salinity, excess	Magnesium, potassium, and/or sodium
Scum production	Die-off of large numbers of bacteria
Septicity	ORP < −100 mV
Shearing action (excess turbulence)	High rate of aeration
Slug discharge of soluble cBOD	3× typical quantity of soluble cBOD
Surfactant	Anionic detergent
Total dissolved solids (TDS), excess	>5000 mg/L
Toxicity	Over-chlorination of MLVSS
Undesired filamentous organism growth	>5 filamentous organisms per floc particle
Viscous floc or Zoogloeal growth	Rapid proliferation of floc-forming bacteria
Low MCRT	<3 days MCRT

- Low dissolved oxygen concentration
- Low pH/High pH
- Nutrient deficiency
- Salinity
- Scum production
- Septicity
- Shearing action (excess turbulence)
- Slug discharge of soluble cBOD
- Surfactants
- Total dissolved solids (TDS)
- Toxicity
- Undesired filamentous organism growth
- Viscous floc or Zoogloeal growth
- Low mean cell residence time (MCRT)

COLLOIDS

Colloids do not dissolve in wastewater and do not settle in wastewater. Colloids also are water-retentive. The adsorption of colloids to floc particles results in the loss of density of the floc particles and restriction of water through the floc particle.

Floc particles possess openings or channels (Figure 11.3). Therefore, when floc particles settle, water must pass around and through the floc particles. The presence

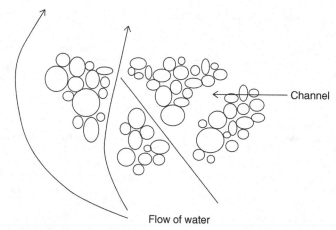

Figure 11.3 *Cross-sectional view of floc particle with channels. A cross-sectional view of a floc particle reveals the presence of numerous openings or channels. These channels permit the flow of substrate, nutrients, dissolved oxygen, and nitrate to the core of the floc particle. When the floc particle settles during the Settle Phase, water must move not only around the floc particle but also through the floc particle.*

of colloids and water molecules adsorbed to colloids hinders the movement of water through the floc particles. The presence of water adsorbed or "bonded" to floc particles is known as "hydrous" floc or "water-bound" floc.

Although colloids are a component of fecal waste, fecal colloids are not excessive in quantity and do not contribute significantly to settleability problems. However, if an SBR receives wastes that are excessive in colloids such as diary wastewater and slaughterhouse wastewater, settleability problems and loss of solids may be experienced due to the adsorption of colloids to floc particles.

ELEVATED TEMPERATURES

With increasing wastewater temperature, treatment efficiency increases with respect to the degradation of cBOD and nBOD. This is due to an increase in biological activity (Figure 11.4). However, there is an optimal value or range of values at which increasing the temperatures no longer provides increased degradation of BOD, but actually is detrimental to treatment efficiency. For small wastewater treatment systems, warm wastewater from industrial discharges, discharges of hot fats, oils, and grease, and discharges of laundry wastewater from commercial establishments are of concern.

The optimum temperature range for the degradation of nBOD or nitrification is approximately 30 °C to 32 °C, while the optimum temperature for the degradation of cBOD is approximately 35 °C. At temperature values of >35 °C, several undesired impacts with respect to operational conditions may occur. First, enzymes, the "tools" that bacteria use for degrading BOD, become denatured or damaged. This results in decreased treatment efficiency. Second, ciliated protozoa numbers and activity decrease. This results in inefficient removal of "fine" solids from the bulk solution due to the lack of consumption of dispersed bacteria by ciliated protozoa and lack of adsorption of particulate material and colloids to floc particles (Figure 11.5).

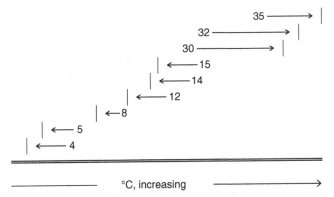

Figure 11.4 *Impact of wastewater temperature upon wastewater treatment. There are several critical temperatures that impact wastewater treatment efficiency. These temperatures are 4, 5, 8, 12, 14, 15, 30, 32, and 35 degrees Celsius. At 4 °C and lower, significant loss of bacterial and protozoan activity occurs. At 5 °C and lower, loss of denitrification occurs. At 8 °C and lower, loss of nitrification occurs, and the proliferation of the foam-producing filamentous organism, Microthrix parvicella, is favored at temperatures as low as 8 °C. At 12 °C and lower, floc formation is inhibited and the production of mature floc particles may take several weeks to develop. At 14 °C and lower, the loss of approximately 50% of the ability of the SBR to degraded cBOD occurs unless appropriate operational measures are implemented. At 15 °C and lower, the loss of approximately 50% of the ability of the SBR to nitrify occurs unless appropriate operational measures are implemented. At 30 °C, ammonia-oxidizing bacteria (AOB) are active at their optimum temperature value. At 32 °C and higher, significant loss of protozoan activity and numbers occurs. Also, the production of firm and dense floc particles is inhibited. At 35 °C, mesophilic methane-forming bacteria are active at their optimum temperature value.*

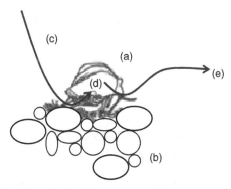

Figure 11.5 *Ciliated protozoa cropping and coating action. Ciliated protozoa perform two significant roles, cropping action and coating action, in the treatment of wastewater in an SBR. First, a ciliated protozoa such as a crawling (creeping) protozoa (**a**) on a floc particle (**b**) creates a water current (**c**) by the beating action of its cilia its ventral (belly) surface that is in contact with the floc particle. As the water current moves between the crawling protozoa and the floc particle, the protozoa consumes dispersed bacterial cells in the water current (**d**). This is known as cropping action. In addition, secretions (**e**) from the crawling protozoa that are released to the bulk solution coat the surface of dispersed cells, colloids, and particulate material (fine solids). This is known as coating action. Once dispersed cells, colloids, and particulate material have been coated, their surface charge is changed and made amenable for adsorption to floc particles. Cropping action and coating action help to cleanse the bulk solution of fine solids.*

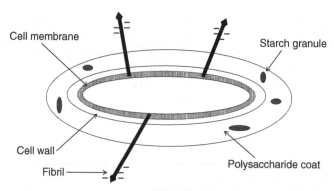

Figure 11.6 *Cellular components necessary for floc formation. There are three significant cellular components that are necessary for floc formation. These components consist of a polysaccharide coat, starch granules such as β-hydroxybuytrate, and fibrils. The polysaccharide coat acts as a "glue" that sticks the bacterial cells together. The starch granules that are located in the polysaccharide coat help to anchor the cells together. The fibrils are an extension of the cell membrane and extend through the cell wall and polysaccharide coat into the bulk solution. The fibrils are approximately 2–5 nm in size and possess a net negative charge. Fibrils from different bacteria join together by bonding with calcium ions (Ca^{2+}) in the bulk solution.*

Third, a weakening of floc particle strength may occur due to the inability of bacterial cells to produce the necessary cellular components required for floc formation (Figure 11.6). These undesired impacts result in the loss of solids and BOD in the decant.

FOAM PRODUCTION

Foam consists of entrapped air or gas bubbles beneath a thin layer of solids or biological secretions. There are several operational conditions that are responsible for foam production and are reviewed in Chapter 13, entitled "Troubleshooting Foam and Scum Production." The production of foam in an SBR often results in the discharge of solids in the decant.

INCREASE IN PERCENT MIXED-LIQUOR VOLATILE SUSPENDED SOLIDS (MLVSS)

The volatile content of the solids or mixed-liquor volatile suspended solids (MLVSS) consists mostly of bacteria. Protozoa make up approximately 5% of the volatile content, while inert material including fats, oils, and grease, biological secretions, colloids, and particulate material contribute to volatile content.

With increasing volatile content of the inert material in the floc particle, the MLVSS becomes less dense and the channels in the floc particles become constricted or plugged (Figure 11.7). With solids being less dense and the flow of water through the channels being reduced or prevented, the floc particles float more easily and may be lost from the SBR during the Decant Phase.

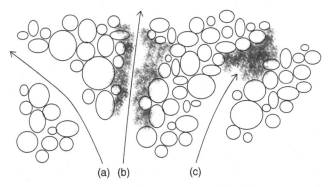

(a) (b)   (c)

Figure 11.7 *Constriction and plugging of channels in floc particles. In a cross-sectional view of a floc particle, there are openings or channels that permit the movement of water through the floc particle as the particle settles during Settle Phase. If a channel experiences no significant build up of wastes* **(a)**, *the water flows easily through the floc particle and settleability of the particle is not hindered. If a channel experiences a build up of wastes* **(b)**—*that is, become constricted—the water flows slowly through the channel and settleability of the particle is hindered. If the channel experiences a significant build up of wastes* **(c)**, *the channel becomes plugged and water cannot pass through the channel and settleability of the particle is severely hindered. Also, a plugged channel may capture air and gas bubbles, further hindering settleability of the particle. The entrapped air and gas bubbles contribute to the production of foam. Wastes that can accumulate in the channels include stored food under a nutrient deficiency, presence of fats, oils, and grease, accumulation of lipids from foam-producing filamentous organisms, gelatinous products from the rapid growth of Zoogloeal organisms, and copious deposits of polysaccharides from rapid, young bacterial growth during a slug discharge of soluble cBOD.*

Percent volatile content in conventional activated sludge processes including the SBR typically is 70% to 75%, while volatile content in extended aeration processes typically is 60% to 65%. The volatile content of any biological treatment process increases dramatically over a short period of time when excessive fats, oils, and grease are adsorbed to floc particles or in the presence of a nutrient deficiency or slug discharge of soluble cBOD.

LACK OF CILIATED PROTOZOA

Ciliated protozoa (Figure 11.8) perform two critical roles that help to remove "fine" solids from the waste stream. Fine solids consist of dispersed growth, colloids, and particulate material. First, the protozoa consume dispersed growth. The consumption of dispersed growth is referred to as "cropping" action. Second, the protozoa excrete cellular wastes that coat the surface of dispersed growth, colloids, and particulate material. This coating action changes the surface charge of the solids and permits their adsorption to floc particles. Once adsorbed, the solids are removed from the waste stream.

A lack of ciliated protozoa (<100 per milliliter) results in a decrease in cropping action and a decrease in coating action. A lack of ciliated protozoa results in an increase in fine solids or TSS in the decant.

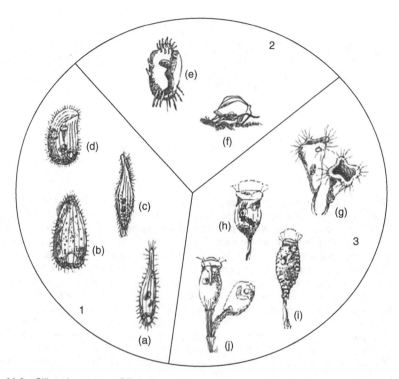

Figure 11.8 *Ciliated protozoa. Ciliated protozoa possess rows of short hair-like structures, cilia, on the surface of their body. The cilia beat to provide locomotion and produce a water current that delivers bacteria to the mouth opening. Ciliated protozoa perform roles in the wastewater treatment process. These roles include cropping action (consumption of dispersed cells), coating action (removal of fine solids through their adsorption to floc particles), and bioindicators of the "health" of the biomass. There are three groups of ciliated protozoa. These groups are (1) the free-swimming ciliates that have cilia on the entire surface of the body and swim in the bulk solution, (2) crawling or creeping ciliates that have cilia only on the ventral or "belly" surface of the body and swim or "crawl" on the surface of floc particles, and (3) stalk ciliates that have cilia in a row surrounding the mouth and possess an enlarged anterior portion or "head" and a slender posterior portion or "tail." Examples of free-swimming ciliates include* Trachelophyllum pusillum *(a),* Tetrahymena pyriformis *(b),* Litonotus fasciola *(c), and* Chilodonella uncinata *(d). Examples of crawling ciliates include* Stylonchia pustulata *(e) and* Aspidisca costata *(f). Examples of stalk ciliates include* Tokophrya quadripartite *(g),* Vorticella alba *(h),* Vorticella striata *(i), and* Epistylis rotans *(j).*

There are several operational conditions that permit a relatively spare population or lack of ciliated protozoa in an SBR. These conditions include (1) start-up, (2) low MCRT, (3) low dissolved oxygen level, (4) wash-out, (5) toxicity or recovery from toxicity, and (6) presence of complex (industrial) waste as the major influent cBOD (substrate) for the bacterial community.

LOW DISSOLVED OXYGEN CONCENTRATION

Minimum dissolved oxygen concentrations are required during React Phase to provide for endogenous respiration, floc formation, and nitrification and to prevent the growth of low dissolved oxygen filamentous organisms (Table 11.2).

TABLE 11.2 Dissolved Oxygen Requirements during React Phase

Operational Requirement	Minimum Dissolved Oxygen (mg/L)
Endogenous respiration	0.8
Floc formation	1
Nitrification	2–3
Prevent undesired growth of low dissolved oxygen filamentous organisms	Increasing with increasing COD removed

A low dissolved oxygen concentration (<0.8 mg/L for 10 consecutive hours) permits significant increase in fine solids in the decant. First, decreased ciliated protozoan activity occurs as well as a decrease in the number of ciliated protozoa. These conditions reduce cropping action and coating action that are helpful in removing fine solids. Second, a low dissolved oxygen concentration hinders the production of cellular components that are necessary for floc formation. This results in the production of weak and buoyant floc particles that are easily sheared and float in the supernatant during Settle Phase. Third, a low dissolved oxygen concentration contributes to the rapid and undesired growth of several filamentous organisms, including *Haliscomenobacter hydrossis*, *Microthrix parvicella*, *Sphaerotilus natans*, and type 1701. In significant numbers these organisms hinder solids settleability. All but *Microthrix parvicella* can be controlled with the use of the Mix Fill Phase.

LOW pH/HIGH pH

Low pH values (<6.5) and high pH values (>8.5) have several adverse impacts upon the SBR, including the interruption of floc formation. The interruption of floc formation occurs when the pH of the SBR causes (1) damage to cellular components that are necessary for floc formation or (2) inhibition of the development of cellular components that are necessary for floc formation. The occurrence of a low pH or high pH in an SBR may be biological or chemical in origin and is reviewed in Chapter 12, entitled "Troubleshooting Undesired Change in pH and Alkalinity."

NUTRIENT DEFICIENCY

A nutrient deficiency (usually for nitrogen or phosphorus) results in the storage of large quantities of insoluble starches in floc particles. The stored food is less dense than wastewater and often "plugs" the channels in the floc particles that permit the flow of water through the floc particles. Once the channels are plugged, air and gas bubbles often are captured in the channels. The captured air and gas bubbles increase the buoyant nature of the floc particles.

The presence of stored food, restriction of water flow through the channels and presence of air and gas bubbles results in poor settleability of floc particles or solids. Poorly settling floc particles contribute to the presence of significant TSS in the decant. The occurrence of a nutrient deficiency is reviewed in Chapter 13, entitled "Troubleshooting Foam and Scum Production."

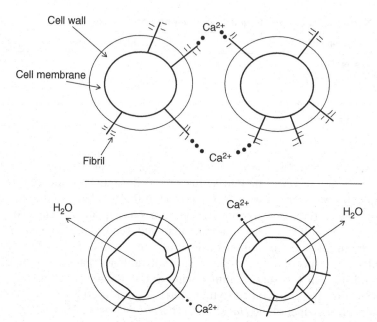

Figure 11.9 *Structural damage to cell membrane caused by excess salinity. Under acceptable salinity concentrations (<5000 mg/L) the cell membrane presses tightly to the inside of the cell wall (**top**). The fibrils that extend from the cell membrane through the cell wall to the bulk solution extend an adequate length to provide for floc formation. In the presence of excess salinity (K^+, Mg^{2+}, and/or Na^+) or excess total dissolved solids (**bottom**), water within the cytoplasm or "gut" content of the cell that is surrounded by the cell membrane leaves the cell to the bulk solution. When this happens, the cell membrane pulls away from the cell wall, as it pulls away, the fibrils are pulled from the bulk solution. Damage to the loss of intracellular water, damage to cell membrane, and withdrawal of fibrils from the bulk solution are responsible for loss of floc formation and die-off of large numbers of bacteria.*

SALINITY

Salinity consists of three metal ions, magnesium (Mg^{2+}), potassium (K^+), and sodium (Na^+). The concentration of these metal ions individually or collectively affect the integrity of the cell membrane of bacteria, bonding of fibrils, and the ability to regulate the flow of materials into and out of the cell (Figure 11.9). If the integrity of the cell membrane is damaged by a high salinity value (>5000 mg/L), floc formation is adversely affected.

Wastewater from ion exchange units and water softeners may contribute to an increase in salinity. Industrial wastewater, such as pickle processing, may also contribute to an increase in salinity. Frac water from gas drilling operations that is discharged to an SBR contributes to an increase in salinity.

SCUM PRODUCTION

Scum in an activated sludge process is brown "flaky" soap. Scum production occurs when a large number of bacteria die over a relatively short period of time due to toxicity or lack of adequate food.

When bacteria die, they undergo autolysis ("self-splitting"). Autolysis results in the release of intracellular chemical components to the bulk solution. Many chemicals are fatty acids. When the acids combine with calcium ions (Ca^{2+}) in the bulk solution, an insoluble soap is produced. Because the soap is produced in brown mixed liquor, the soap is brown and is easily observed on the surface of an SBR during Settle Phase and Decant Phase. The soap also may be observed on the surface of foam. The occurrence of scum production is reviewed in Chapter 13, entitled "Troubleshooting Foam and Scum Production."

SEPTICITY

Septicity is the degradation of cBOD in the absence of free molecular oxygen (O_2) and nitrate (NO_3^-). Malodor production is associated with septicity. Septicity may be defined as the occurrence of an oxidation-reduction potential (ORP or redox) of $\leq -100\,mV$.

Septicity occurs when solids remain for too long a period of time in a sewer system or treatment tank and free molecular oxygen and nitrate are no longer available to bacterial cells for the degradation of cBOD. Septicity and malodor production begin 30–180 minutes after free molecular oxygen and nitrate are no longer available. Hydrogen sulfide (H_2S) usually is produced during septicity, along with many malodorous, volatile fatty acids, volatile nitrogen-containing compounds, and volatile sulfur-containing compounds (Table 11.3).

SHEARING ACTION

Shearing action occurs in the presence of excess turbulence and may be caused by high rates of aeration and mixing. Shearing action rips floc particles and results in the release of fine solids. A good indicator of excess turbulence in the mixed liquor is the microscopic detection of excessive dispersed growth (Figure 11.10).

TABLE 11.3   Examples of Malodorous Compounds Produced under Septicity

Compound	Formula	Comment
Acetate	CH_3COOH	Fatty acid
Butyrate	$CH_3(CH_2)_2COOH$	Fatty acid
Formate	$HCOOH$	Fatty acid
Propionate	CH_3CH_2COOH	Fatty acid
Cadaverine	$H2N(CH_2)_5NH_2$	Nitrogen-containing
Methylamine	CH_3NH_2	Nitrogen-containing
Putresine	$H_2N(CH_2)_4NH2$	Nitrogen-containing
Skatole	$CH_3NCH_3CH_3$	Nitrogen-containing
Benzyl mercaptan	$C_6H_5CH_2SH$	Sulfur-containing
Ethyl mercaptan	C_2H_5SH	Sulfur-containing
Methyl mercaptan	CH_3SH	Sulfur-containing

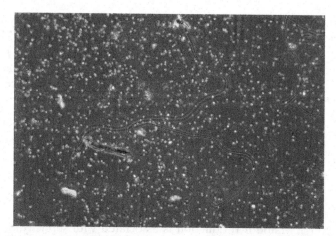

Figure 11.10   *Excessive dispersed growth. Dispersed growth consists of small (<10μm) and spherical floc particles. In an "unhealthy" biomass the relative abundance of dispersed growth may be "excessive" or "hundreds" (100, 200, 300, ...) of small floc particles or "cells" per field of view at 100× total magnification.*

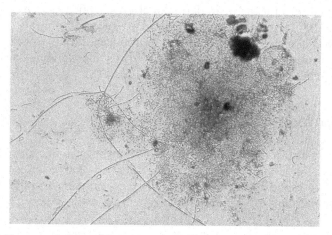

Figure 11.11   *Young bacterial growth from a slug discharge of soluble cBOD. In the presence of adequate dissolved oxygen and adequate nutrients, bacterial cells rapidly degrade a slug discharge of soluble cBOD. The rapid degradation results in the proliferation of young bacterial growth. Young bacterial cells produce a copious quantity of polysaccharides that separate the cells a great distance as compared to old bacterial cells. In the Gram-stained smear of mixed liquor, the floc particle contains a core of old bacteria that are tightly compacted and a perimeter of young bacteria that are loosely compacted due to the relatively large quantity of polysaccharides.*

SLUG DISCHARGE OF SOLUBLE CBOD

A slug discharge of soluble cBOD is considered to be a quantity of cBOD that is 2–3 times greater than the typical quantity of cBOD and is received over a 2–4-hour period. Slug discharges usually are associated with industrial wastewater, but the rapid transfer or discharge of septic wastewater (recycle stream) to an SBR can mimic a slug discharge.

A slug discharge promotes rapid, young bacterial growth (Figure 11.11). This growth is associated with the production of a copious quantity of insoluble polysaccharides. The polysaccharides promote the development of weak and buoyant floc particles that are easily sheared and contribute to excess TSS in the decant.

The polysaccharides also permit the development of buoyant floc particles that contribute to an increase in distance between agglutinated bacterial cells (loss of density) and capture air and gas bubbles.

SURFACTANTS

Surfactants, especially anionic surfactants, contribute to four operational problems. These problems are (1) increase in size of air bubbles, (2) dispersion of floc particles and loss of fine solids, (3) foam production, and (4) toxicity.

As surface active agents, surfactants change the size of air bubbles. Depending on the quantity of surfactants in an SBR and the increase in size of air bubbles, oxygen transfer efficiency may be reduced. Because they are surface-active agents, surfactants also produce billowy white foam.

Because floc particles contain billions of bacteria that possess a net negative charge, anionic surfactants easily disperse floc particles. Dispersed floc particles are easily sheared during the React Phase and often float during the Settle Phase.

Surfactants also are toxic or may contribute to toxicity. The toxicity of surfactants is influenced by several operational factors, including dissolved oxygen concentration, hardness of the wastewater, and the chemical structure of the surfactant (Figure 11.12) of the surfactant. Although some surfactants are not toxic, they can contribute to toxicity by increasing the pore size in the bacterial cell wall and cell membrane and permit some toxic wastes to enter the cell that normally would not in the absence of the surfactant. The occurrence of surfactant foam is reviewed in Chapter 13, entitled "Troubleshooting Foam and Scum Production."

TOXICITY

Toxicity results in the death of bacteria and the interruption of floc formation. The death of bacteria contributes to the loss of many intracellular components, especially colloids and particulate material to the bulk solution.

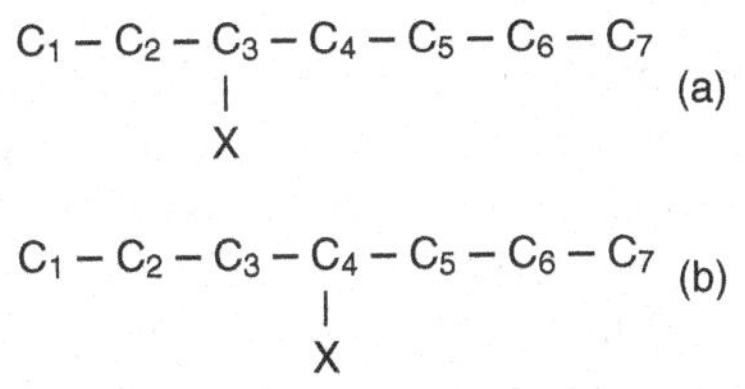

$$C_1 - C_2 - C_3 - C_4 - C_5 - C_6 - C_7$$
$$|$$
$$X$$
(a)

$$C_1 - C_2 - C_3 - C_4 - C_5 - C_6 - C_7$$
$$|$$
$$X$$
(b)

Figure 11.12 *Chemical structures in surfactants. Although there are numerous types of chemical structures for surfactants, two surfactants with the same chemical formula, net charge, and molecular formula may differ greatly with respect to their toxic nature in the activated sludge process. For example, two surfactants (a and b) have the same chemical formula (C_7X_1) and same chemical structure (aliphatic with single bonds only) but have a different position for the attachment of the "X" group. In figure a the "X" group is on the third carbon atom, while in figure b the "X" group is on the fourth carbon atom. This slight change in position of the "X" group may cause one surfactant—for example a,—to be toxic, while the other surfactant, b, is nontoxic. Changes in chemical structures in surfactants may cause the treatment process to experience toxicity on one day and no toxicity on another day, even though the same quantity of the surfactant is received at the treatment process.*

Toxicity may be detected by noting a significant decrease in the specific oxygen uptake rate (SOUR) during the React Phase. Toxicity may occur from inorganic ions or compounds or organic compounds. Some inorganic ions that can contribute to toxicity include heavy metals, and some inorganic compounds that can contribute to toxicity include hydrogen sulfide (H_2S) and high concentrations of ammonia (NH_3). Organic compounds that can contribute to toxicity include cleaning agents, disinfectants, solvents, and surfactants.

A zero SOUR value is indicative of a total kill in the mixed liquor and requires the identification and regulation of the toxic discharge. Reseeding of the SBR slowly with influent wastewater or quickly with bioaugmentation products or mixed liquor from a "healthy," activated sludge process is required. A low SOUR value may indicate a change in composition of the BOD, a decrease in strength of the BOD, or inhibition or partial kill in the mixed liquor. For inhibition or partial kill, an increase in the number of "healthy" and active bacteria is required. Again, reseeding with bioaugmentation products or mixed liquor from a "healthy," activated sludge process can provide for a rapid recovery.

TOTAL DISSOLVED SOLIDS (TDS)

Total dissolved solids (TDS) impact floc formation in similar fashion as salinity. TDS has an adverse impact upon floc formation at approximately 5000 mg/L. Wastewater from ion exchange units, water softeners, food processing firms, and industries contribute to increases in TDS.

UNDESIRED FILAMENTOUS ORGANISM GROWTH

Filamentous organisms perform positive and negative roles in the activated sludge process. Positive roles include (1) the degradation of cBOD, (2) the degradation of some complex forms of cBOD, and (3) the strengthening of floc particles. Negative roles include (1) loss of settleability, (2) loss of solids, and (3) production of viscous chocolate-brown foam. Foam production is associated with a limited number of filamentous bacteria, including Nocardioforms, *Microthrix parvicella*, and type 1863.

The occurrence of negative roles begins when filamentous organisms are present in undesired numbers (Table 11.4). Undesired filamentous organism growth is considered to be the presence of more than five filamentous organisms in most floc particles. In order to determine the relative abundance of filamentous organisms per floc particle, a microscopic examination of a wet mount of mixed liquor or Gram-stained smear of mixed liquor should be examined at 100× total magnification.

Filamentous organisms outgrow floc bacteria for several operational conditions (Table 11.5). Common operational conditions that permit the undesired growth of filamentous organisms include (1) excess fats, oils and grease, (2) high MCRT, (3) low dissolved oxygen level, (4) low F/M, (5) nutrient deficiency, (6) readily-degradable substrates such as short chain acids and alcohols, and (7) septicity and sulfides.

TABLE 11.4 Relative Abundance of Filamentous Organisms

Rating	Description
0	Filamentous organisms not observed
1	Filamentous organisms present, but found in an occasional floc particle in very few fields of view
2	Filamentous organisms present, but found only in some fields of view
3	Filamentous organisms observed in most floc particles at low density (1–5 filamentous organisms per floc particle)
4	Filamentous organisms observed in most floc particles at medium density (6–20 filamentous organisms per floc particle)
5	Filamentous organisms observed in most floc particles at high density (>20 filamentous organisms per floc particle)
6	Filamentous organisms observed in most floc particles; filamentous organisms more abundant than floc particles; or filamentous organisms growing in large numbers in the bulk solution

TABLE 11.5 Operational Conditions Associated with the Undesired Growth of Filamentous Organisms

Operational Condition	Filamentous Organism
High MCRT (>10 days)	0041, 0092, 0581, 0675, 0803, 0961, 1851, *Microthrix parvicella*
Fats, oils, and grease	0092, *Microthrix parvicella*, Nocardioforms
High F/M or slug discharge of soluble cBOD	1863
High pH (>7.4)	*Microthrix parvicella*
Low dissolved oxygen and high MCRT	*Microthrix parvicella*
Low dissolved oxygen and low MCRT	*Haliscomenobacter hydrossis*, *Sphaerotilus natans*, 1701
Low F/M (<0.05)	*Haliscomenobacter hydrossis*, *Microthrix parvicella*, Nocardioforms, 0041, 0092, 0581, 0675, 0803, 0961, 021N
Low nitrogen or phosphorus	*Haliscomenobacter hydrossis*, Fungi, Nocardioforms, *Sphaerotilus natans*, *Thiothrix*, 0041, 0092, 0675, 1701, 021N
Low pH (<6.5)	Fungi, Nocardioforms
Organic acids	*Beggiatoa*, *Thiothrix*, 021N
Readily degradable substrates	*Haliscomenobacter hydrossis*, *Nosticoda limicola*, *Sphaerotilus natans*, *Thiothrix*, 1851, 021N
Septicity/sulfides (1–15 mg/L)	*Beggiatoa*, *Nosticoda limicola*, *Thiothrix*, 0041, 021N
Slowly degradable substrate	*Microthrix parvicella*, Nocardioforms, 0041, 0092
Warm wastewater temperature	*Sphaerotilus natans*, 1701
Winter proliferation	*Microthrix parvicella*

VISCOUS FLOC OR ZOOGLOEAL GROWTH

Viscous floc or Zoogloeal growth is the rapid and undesired growth of floc-forming bacteria in the amorphous form (Figure 11.13) or dendritic or tooth-like form (Figure 11.14). The growth results in the production of "slimy" or viscous floc particles. The viscous nature is due to the cellular secretion of a copious quantity of gelatinous material. Viscous floc is weak and buoyant and easily sheared. Its weak and buoyant composition contributes to numerous solids in the decant. The gelatinous material captures air and gas bubbles and often produces billowy white foam.

Operational conditions that contribute to Zoogloeal growth include (1) a septic or anaerobic/fermentative condition upstream of an aerated condition, (2) nutrient deficiency, (3) high MCRT, and (4) significant swings in F/M. Treatment processes that practice biological phosphorus removal typically have significant Zoogloeal growth. The growth can be controlled with the use of anoxic periods or Mix Fill Phase.

Viscous floc is known as Zoogloeal growth. The term "Zoogloeal" was obtained from the genus name of the bacterium *Zoogloea ramigera*. This bacterium was the first floc-forming bacterium that was observed and identified as growing rapidly in the mixed liquor and producing buoyant and weak floc particles.

LOW MCRT

Low MCRT is considered to be <3 days. Young sludge or young bacteria produce relatively large quantities of polysaccharides that are insoluble in wastewater and

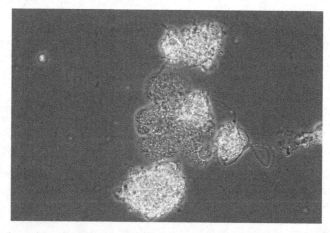

Figure 11.13 *Amorphous Zoogloeal growth. Zoogloeal growth is the rapid and undesired growth of floc-forming bacteria. This growth results in the production of weak and buoyant floc particles due to the separation of the bacteria by a copious quantity of gelatinous material. Amorphous Zoogloeal growth (growth without specific form or shape) is most commonly observed in activated sludge process and can be seen in the center of the photomicrograph as poorly compacted or loosely aggregated globular masses of bacterial cells.*

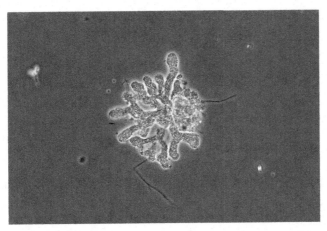

Figure 11.14 *Dendritic Zoogloeal growth. Dendritic ("tooth-like" or "finger-like") Zoogloeal growth is less often observed in activated sludge processes as amorphous Zoogloeal growth. Dendritic usually appears at young sludge ages or mean cell residence times (MCRT).*

less dense than wastewater. The polysaccharides also contribute to a decrease in the density of the floc particles. The presence of polysaccharides and loss of compaction of bacterial cells permits the formation of weak and buoyant floc particles that are easily sheared and float.

A young sludge should occur only once in an SBR. This is during start-up. However, a young sludge can develop through (1) recovery from toxicity, (2) loss of significant solids in the decant, and (3) over-wasting of solids.

TROUBLESHOOTING KEY FOR THE IDENTIFICATION OF OPERATIONAL FACTORS RESPONSIBLE FOR HIGH DECANT TSS

1. Has the SBR experienced a significant discharge of colloids—for example, dairy or slaughterhouse wastewater?

   Yes   See 2
   No   See 3

2. Identify and regulate or pretreat the discharge of colloids. The addition of bio-augmentation products to the SBR or colloid-containing waste stream may help to reduce the undesired impact of colloids on the SBR.

3. Has foam or scum production occurred?

   Yes   See 4
   No   See 5

4. Identify the foam or scum and implement appropriate corrective measures. See Chapter 13, entitled "Troubleshooting Foam and Scum Production."

5. Has a significant increase in percent MLVSS occurred over a relatively short period of time?

Yes   See 6
No   See 7

6. An increase in percent MLVSS may be due to the discharge of fats, oils, and grease or the biological production of starches under a nutrient deficiency, slug discharge of soluble cBOD or Zoogloeal growth. Determine the causative factor for the increase in percent MLVSS and correct accordingly.

7. Is there a lack of ciliated protozoa, that is, <100 per milliliter in the mixed liquor?

Yes   See 8
No   See 9

8. A lack of ciliated protozoa is typically an indicator of an adverse operational condition such as start-up, low dissolved oxygen concentration, toxicity, recovery from toxicity, or presence of complex wastes. Determine the causative factor for the lack of ciliated protozoa and correct accordingly.

9. Has a significant change in pH occurred—for example, a significant decrease or increase in mixed liquor pH?

Yes   See 10
No   See 11

10. The factors responsible for the change in pH and its correction can be identified in Chapter 12, entitled "Troubleshooting Undesired Change in Alkalinity and pH."

11. Has a nutrient deficiency occurred in the SBR?

Yes   See 12
No   See 13

12. The factors responsible for a nutrient deficiency and its correction can be identified in Chapter 13, entitled "Troubleshooting Foam and Scum Production."

13. Has septicity occurred in the sewer system or lift station serving the SBR or has a septic recycle stream been discharged to the SBR?

Yes   See 14
No   See 15

14. Septicity can be detected by an ORP value $\leq -100\,mV$, a decrease in alkalinity and pH, and production of malodor. Septicity may be corrected by (1) increasing aeration (dissolved oxygen concentration), (2) cleaning and disinfecting the sewer system or lift station, (3) adding sodium nitrate or calcium nitrate, (4) pH adjustment, (5) sulfide precipitation, or (6) adding appropriate bioaugmentation products.

15. Has shearing action occurred in the SBR?

Yes   See 16
No   See 17

16. Shearing action may be due to excess aeration or mixing. Check current aeration rate and mixing rate and, if possible, reduce the rate(s) to decrease shearing action. If aeration rate has been increased to provide more dissolved oxygen, the dissolved oxygen requirement may be atypical for the SBR. Coagulants (metal salts) and/or polymers may be added to the SBR to capture sheared solids.

17. Has a slug discharge of soluble cBOD occurred?

Yes   See 18
No   See 19

18. The factors responsible for a slug discharge of soluble cBOD can be identified in Chapter 13, entitled "Troubleshooting Foam and Scum Production."

19. Have excess surfactants been discharged to the SBR?

Yes   See 20
No   See 21

20. For SBR that experience occasional problems or suspect occasional problems due to excess surfactants, periodic monitoring of influent surfactants should be performed. Testing for the concentration of anionic surfactants (typically the most problematic surfactants for activated sludge process) is performed with the MBAS (methylene blue active substance) test.

21. Has toxicity occurred in the SBR?

Yes   See 22
No   See 23

22. Toxicity in an SBR may be detected by performing a specific oxygen uptake rate (SOUR). If toxicity has occurred, the source of toxic waste should be identified and regulated. To quickly increase the number of active bacteria in the SBR, bioaugmentation products may be added.

23. Has undesired filamentous organism growth occurred?

Yes   See 24
No   See 25

24. Undesired filamentous organism growth must be determined by a microscopic examination of a wet mount of mixed liquor or a Gram-stained smear of mixed liquor. With increasing numbers of filamentous organisms—for example, >5 filamentous organisms per floc particle—poor compaction of solids and loss of solids from the SBR may occur. Commonly used techniques for controlling undesired filamentous organism growth include wasting of solids and chlorination of solids. Control of some filamentous organism growth may be obtained by using Static Fill Phase or Mix Fill Phase. Improved settleability also may be achieved by adding a coagulant (metal salt) or polymer to the mixed liquor during the React Phase.

25. Has the development of viscous floc or Zoogloeal growth occurred?

Yes   See 26
No   See 27

26. Viscous floc can only be confirmed with a microscopic examination of a wet mount of mixed liquor. Viscous floc can be controlled with the use of Mix Fill Phase. Chlorination of the mixed liquor destroys viscous floc but results in the production of much dispersed growth. Operational conditions, especially upstream septicity and nutrient deficiency that promote the growth of viscous floc, should be identified and corrected.

27. Is the SBR experiencing a condition that mimics a low MCRT—that is, have too many solids been removed from the SBR or is the system recovering from toxicity? Identify the condition and correct it.

Check List for the Identification of Operational Factors Responsible for High Decant TSS

Operational Factor	√ If Monitored	√ If Possible
Colloids		
Elevated temperature		
Foam production		
Increase in percent MLVSS		
Lack of ciliated protozoa		
Low dissolved oxygen concentration		
Low pH/high pH		
Nutrient deficiency		
Salinity		
Scum production		
Septicity		
Shearing action		
Slug discharge of soluble cBOD		
Surfactants		
Total dissolved solids		
Toxicity		
Undesired filamentous organism growth		
Viscous floc or Zoogloeal growth		
Young sludge age or low MCRT		

12

Troubleshooting Undesired Changes in pH and Alkalinity

BACKGROUND

For most bacteria the optimum pH for activity and reproduction is near neutral (pH 7), and little activity and reproduction occur at values of ±1 unit of their optimum pH. Few bacteria are active and reproduce at pH values of <4 and >9.5. Most biological treatment systems, including SBR, operate at pH values near neutral (6.8–7.2) to promote acceptable bacterial activity and reproduction. Biological treatment systems may experience operational problems below or above a near-neutral pH value.

Operational problems that may occur in SBR that experience pH values of <6.8 include the following:

- Decrease in enzymatic activity (loss of treatment efficiency)
- Increase in hydrogen sulfide (H_2S) production
- Inhibition of nitrification
- Interruption of floc formation
- Undesired growth of filamentous fungi and some Nocardioforms

Operational problems that may occur in SBR that experience pH values of >7.2 include the following:

Troubleshooting the Sequencing Batch Reactor, by Michael H. Gerardi
Copyright © 2010 by John Wiley & Sons, Inc.

- Decrease in enzymatic activity (loss of treatment efficiency)
- Increase in conversion of ammonium (NH_4^+) to ammonia (NH_3)
- Interruption of floc formation
- Increase in the growth of the filamentous organism *Microthrix parvicella*

Alkalinity refers to the capacity of the wastewater to neutralize acids; thus alkalinity has a relationship with pH. Alkalinity is not the same as pH, because wastewater does not need a high pH to have high alkalinity. The major components of alkalinity consist of carbonates (CO_3^{2-}), bicarbonates (HCO_3^-), and hydroxides (OH^-). Some minor components of alkalinity include phosphates (PO_4^{3-}) and silicates (SiO_4^{4-}). These components act as buffering agents and help to maintain a stable pH. Alkalinity is important in biological wastewater treatment processes. Alkalinity performs the following roles:

- Serves as a pH buffer that helps to maintain a near-neutral pH and proper enzymatic activity
- Serves as a carbon source for nitrifying bacteria and helps to promote nitrification
- May be associated with high concentrations of dissolved solids that interrupt floc formation

The major compounds that provide for alkalinity in wastewater are bicarbonates, carbonates, and hydroxides. In engineering terms, alkalinity often is expressed as mg/L of calcium carbonate ($CaCO_3$).

The relationship between alkalinity and pH is not always direct; that is, an increase or decrease in alkalinity does not always result in an increase or decrease in pH, respectively. Sodium bicarbonate ($NaHCO_3$) or baking soda, for example, produces a rapid and significant increase in alkalinity when added to wastewater but does not produce a significant increase in pH, whereas calcium hydroxide ($Ca(OH)_2$) or hydrated lime produces a significant increase in pH when added to wastewater but does not produce an immediate and significant increase in alkalinity as sodium carbonate produces.

Many surface, potable water supplies that supply the bulk of the alkalinity to wastewater treatment systems and experience acid rain deposition may contain relatively low concentrations of alkalinity. Additionally, industrial wastewater that contains acidic compounds also may destroy significant quantities of alkalinity in the sewer system.

The typical pH range for domestic wastewater is 7.0–7.2, while the typical alkalinity range for domestic wastewater is 100–300 mg/L. An SBR that regularly receives wastewater with high or low pH values should have a neutralization system upstream of the SBR. However, an adjustment of the influent pH to a desired pH value does not necessarily mean that the pH or alkalinity of the SBR will be within that desired range. Change in pH and alkalinity in the SBR should occur slowly to avoid damage (inhibition or toxicity) to the biomass, interruption of floc formation, or decrease in treatment efficiency. Also, if primary clarifiers are used, a change in pH or alkalinity can occur across the clarifier if solids retention time is excessive and an anaerobic condition develops.

Appropriate piping for pH and alkalinity addition should be provided to an influent equalization tank or SBR. Routine monitoring and adjustment for pH and alkalinity also should be performed. Alkalinity addition should be based on the quantity needed at the beginning of Decant Phase—that is, after all major biochemical reactions are complete, especially nitrification. The quantity of alkalinity remaining in the decant should be $\geq 50 \, mg/L$ as $CaCO_3$.

The pH and alkalinity values for an operating SBR are influenced by three operational conditions. These conditions are biological activity, chemicals added to the SBR, and chemicals discharged to the SBR.

BIOLOGICAL ACTIVITY

There are six biological activities that occur in SBR that influence alkalinity and pH. Some activities occur continuously, while others are intermittent. Some activities are major, while some are minor. These activities include:

- Respiratory release of carbon dioxide (CO_2)
- Nitrification
- Denitrification
- Sulfate reduction
- Sulfide oxidation
- Fermentation (mixed acid production)

Aerobic reactions destroy or decrease alkalinity, while anaerobic reactions produce or increase alkalinity (Table 12.1). The rate of change (decrease or increase) in alkalinity provides useful information on the rate of biological activity. For example, when plotted, the rate of alkalinity decrease predicts the rate of nitrification (Figure 12.1), while the rate of alkalinity increase predicts the rate of denitrification (Figure 12.2). Monitoring alkalinity change can provide immediate indicators of biological activity in the SBR. Monitoring can be performed on-site and *in situ*.

RESPIRATORY RELEASE OF CARBON DIOXIDE

When carbonaceous biochemical oxygen demand is degraded, carbon dioxide (CO_2) is produced. When carbon dioxide combines with water (H_2O), carbonic acid

TABLE 12.1 Change in Alkalinity According to Aerobic and Anaerobic Reactions

Biochemical Reaction	Alkalinity Change	
	Decrease	Increase
Respiratory release of CO_2; aerobic (O_2) oxidation of cBOD	X	
Nitrification; aerobic oxidation of NH_4^+ to NO_3^-	X	
Denitrification; anaerobic reduction of NO_3^- to N_2		X
Sulfate reduction; anaerobic reduction of SO_4^{2-} to H_2S/HS^-		X
Sulfide oxidation; aerobic oxidation of HS^- to SO_4^{2-}	X	
Termeation, use of CH_2O to degrade cBOD	X	

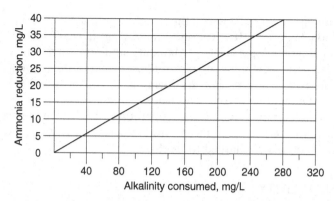

Figure 12.1 *Rate of alkalinity decrease during nitrification. Nitrification results in the loss of alkalinity due to its use by nitrifying bacteria as a carbon source for cellular growth and its destruction by nitrous acid by ammonia-oxidizing bacteria (AOB). The rate of alkalinity decrease (consumption) can be used to monitor the activity of nitrifying bacteria and the success of the SBR to achieve nitrification.*

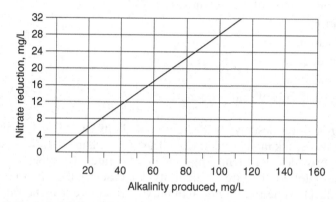

Figure 12.2 *Rate of alkalinity increase during denitrification. Denitrification results in the increase of alkalinity due to the production of hydroxyl ions (OH⁻). The rate of alkalinity increase can be used to monitor the activity of denitrifying bacteria and the success of the SBR to achieve denitrification.*

(H_2CO_3) is produced [Eq. (12.1)]. The production of carbonic acid results in a decrease in pH.

$$CO_2 + H_2O \rightarrow H_2CO_3 \tag{12.1}$$

However, if carbon dioxide is stripped from the SBR by a high aeration rate, then a decrease in pH does not occur. If the pH of the SBR is >7, due to the lack of carbonic acid production, this may have an adverse impact upon the availability of orthophosphate as a phosphorus nutrient for bacterial growth. Stripping of carbon dioxide is common in aerated biological treatment systems, but some systems such as those at high aeration rates or using coarse aeration tend to strip more carbon dioxide than others.

At pH values of <7, most of the orthophosphate exists as $(H_2PO_4^-)$ and is not easily precipitated by calcium ions (Ca^{2+}). At pH values of >7, most of the orthophosphate exists as HPO_4^{2-} and may be easily precipitated by calcium ions. The precipitation of orthophosphate places phosphorus in an insoluble form as calcium phosphate ($CaHPO_4$). In this form, phosphorus is not available as a nutrient for bacterial use.

NITRIFICATION

Nitrification is the biological oxidation of ammonium (NH_4^+) to nitrite (NO_2^-) and then to nitrate (NO_3^-) [Eqs. (12.2) and (12.3)]. Nitrifying bacteria obtain energy from the oxidation of ammonium and nitrite. Nitrifying bacteria obtain carbon for cellular growth (sludge production) from alkalinity. Approximately 7.14 mg/L of alkalinity are consumed per milligram of ammonium oxidized. The loss of alkalinity through nitrification decreases pH. If unchecked, nitrification can produce a significant decrease in alkalinity and pH.

$$NH_4^+ + 1.5O_2 \xrightarrow{\text{Ammonia-oxidizing bacteria}} NO_2^- + H_2O + 2H^+ \quad (12.2)$$

$$NO_2^- + 0.5O_2 \xrightarrow{\text{Nitrite-oxidizing bacteria}} NO_3^- \quad (12.3)$$

DENITRIFICATION

Denitrification is the use of nitrate typically or nitrite atypically by denitrifying bacteria to degrade soluble cBOD [Eq. (12.4)]. In addition to the production of carbon dioxide, water and molecular nitrogen (N_2), hydroxyl ions (OH^-) are produced. Molecular nitrogen is an insoluble gas in wastewater and escapes to the atmosphere. The hydroxyl ions return alkalinity to the wastewater. Approximately 3.6 mg/L of alkalinity are returned to the wastewater for each milligram of nitrate converted to molecular nitrogen. The return of alkalinity through denitrification not only increases alkalinity in the wastewater but also increases the pH of the wastewater.

$$6NO_3^- + 5CH_3OH \xrightarrow{\text{denitrifying bacteria}} 3N_2 + 5CO_2 + 7H_2O + 6OH^- \quad (12.4)$$

SULFATE REDUCTION

Sulfate (SO_4^{2-}) enters an SBR from two major sources. First, it is a component of groundwater and enters through inflow and infiltration (I/I). Second, it is a component of urine and enters through the discharge of domestic wastewater. The typical range of the concentration of sulfate in domestic wastewater is 20–30 mg/L.

In the absence of dissolved oxygen and nitrate, sulfate-reducing bacteria (SRB) use sulfate to degrade soluble cBOD. When soluble cBOD is degraded, sulfide (HS^-) and hydrogen sulfide (H_2S) are produced and alkalinity is returned to the wastewater [Eq. (12.5)]. Approximately 1.04 mg/L of alkalinity is produced for each milligram of sulfate reduced to sulfide/hydrogen sulfide. The return of alkalinity increases pH.

$$cBOD + SO_4^{2-} \xrightarrow{\text{sulfate-reducing bacteria}} H_2O + H_2S/HS^- + OH^- \quad (12.5)$$

Sulfate-reducing bacteria are anaerobes, that is, they cannot use dissolved oxygen or nitrate for the degradation of soluble cBOD. Sulfate-reducing bacteria are active in an anaerobic/fermentative (septic) condition. A septic condition exists when sulfate-reducing bacteria and soluble cBOD are present and dissolved oxygen and nitrate are absent or a gradient for exists for each. In this condition the oxidation–reduction potential (ORP) of the wastewater or sludge is ≤–100 mV. Examples of an ORP ≤–100 mV would occur inside (1) anaerobic digesters, (2) flat sewers with slow wastewater flow, (3) lift stations, (4) treatment tanks with long solids retention time and lack of aeration and mixing, and (5) anaerobic/fermentative zones or periods that are used to control undesired filamentous organism growth and provide for biological phosphorus release.

SULFIDE OXIDATION

Sulfide (HS-) enters an SBR from the anaerobic digestion of soluble cBOD, especially sulfur-containing amino acids and proteins by sulfate-reducing bacteria. These amino acids and proteins contain sulfur (S) or thiol groups (-SH) that are released during degradation and form sulfide and hydrogen sulfide (Figure 12.3). Sulfides may be produced in an SBR during the Static Fill Phase. Here, sulfate is reduced to hydrogen sulfide and/or sulfide.

Reduced sulfur is sulfur that contains hydrogen. It exists as sulfide (HS⁻) and hydrogen sulfide (H₂S). The relative quantity of each form of reduced sulfur is determined by the pH of the wastewater or sludge. At pH values of ≥7, most of the reduced sulfur is in the sulfide form. At pH values of <7, most of the reduced sulfur

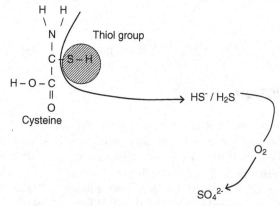

Figure 12.3 *Formation of reduced sulfur from the release of thiol groups (-SH). Thiol groups are found on some amino acids that make of proteins. When these amino acids and proteins are degraded under an anaerobic condition, the released thiol groups form sulfide (HS-) at pH ≥ 7 or hydrogen sulfide at pH < 7. When the thiol groups are released under an aerobic condition, the sulfides/hydrogen sulfides are oxidized biologically and chemically to sulfate (SO₄²⁻).*

is in the hydrogen sulfide form. Hydrogen sulfide is malodorous and toxic. At 1–3 mg/L it is toxic to nitrifying bacteria.

During aeration sulfides are oxidized to sulfate by sulfate-oxidizing bacteria in order to obtain cellular energy [Eq. (12.6)]. Sulfate is converted to sulfuric acid (H_2SO_4). The production of sulfuric acid results in the destruction of alkalinity and a decrease in pH. Depending upon the pH of the SBR, approximately 1.6–3.1 mg/L of alkalinity are destroyed per milligram of sulfide oxidized to sulfuric acid.

$$H_2S + 2O_2 \xrightarrow{\text{sulfide-oxidizing bacteria}} H_2SO_4 \tag{12.6}$$

FERMENTATION

Fermentation or mixed acid production is the degradation of soluble cBOD by fermentative bacteria in the absence of dissolved oxygen and nitrate. Dissolved oxygen and nitrate are molecules that are found outside or external to the bacterial cell and are taken inside the cell to remove (transport) electrons outside the cell when the electrons are freed from the chemical bonds of the degrading soluble cBOD. When dissolved oxygen and nitrate are not available to fermentative bacteria, the absorbed soluble cBOD serves not only as carbon and energy sources for the fermentative bacteria but also as the molecule that transports the freed electrons outside the bacterial cell (Figure 12.4). Numerous short-chain fatty acids

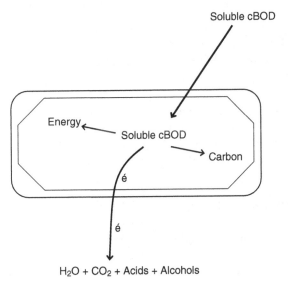

Figure 12.4 *Transport of electrons from bacterial cell through fermentation. In the absence of free molecular oxygen and nitrate, fermentative or acid-forming bacteria absorb soluble cBOD and use the absorb cBOD or another absorbed cBOD as the final electron transport molecule or "wheelbarrow" to transport electrons from the degraded cBOD to the bulk solution. When soluble cBOD enters the cell it is degraded, that is, chemical bonds are broken and electrons are freed from the chemical bonds. The freed electrons surrender some of their energy to the cell and then are carried by the wheelbarrow from the cell to the bulk solution. Products from the fermentation of soluble cBOD include water, carbon dioxide and a mixture of acids and alcohols.*

TABLE 12.2 Examples of Fatty Acids Produced through Fermentation

Fatty Acid	Formula
Acetate	CH_3COOH
Butyrate	$CH_3(CH_2)_2COOH$
Formate	$HCOOH$
Lactate	$CH_3CHOHCOOH$
Propionate	CH_3CH_2COOH

are produced through fermentation (Table 12.2). These acids destroy alkalinity and lower pH.

CHEMICAL COMPOUNDS ADDED TO THE SBR

There are several groups of chemical compounds that may be added to an SBR to improve treatment efficiency. These groups include:

- Chemicals for pH adjustment
- Chemicals for alkalinity addition
- Chemicals for phosphorus removal
- Coagulants or metal salts for solids capture and thickening
- Flocculants
- Disinfectants
- Chemicals for carbon addition for denitrification
- Chemicals for nutrient addition

The addition of chemical compounds to an SBR may produce a gradual or sudden change in pH and/or alkalinity. Therefore, proper calculation of the quantity of a chemical compound to be used, along with thorough mixing and gradual addition of the chemical compound, should be performed.

CHEMICALS FOR pH CONTROL

Chemical compounds may be added to an SBR to maintain a near-neutral pH operating range. The pH of an SBR may need to be increased with the addition of an alkali compound or decreased with the addition of an acidic compound. Chemical compounds commonly used to regulate pH are listed in Table 12.3.

Appropriate safety precautions should be exercised when handling any chemical. The addition of nitric acid and sulfuric acid presents unique biological concerns for an SBR. The addition of nitric acid may contribute to the release of nitrite (NO_2^-) and nitrate (NO_3^-) that represents clumping or denitrification in the Settle Phase and Decant Phase. Also, nitric acid may contribute to an increase in total nitrogen decanted from an SBR. The addition of sulfuric acid may contribute to the

TABLE 12.3 Chemical Compounds Used for pH Control

Chemical Name	Chemical Formula	Common Name	pH Increase	pH Decrease
Calcium carbonate	$CaCO_3$	Limestone	X	
Calcium hydroxide	$Ca(OH)_2$	Lime	X	
Calcium oxide	CaO	Lime	X	
Hydrochloric acid	HCl	Muracic acid		X
Magnesium hydroxide	$Mg(OH)_2$	Mag	X	
Magnesium oxide	MgO		X	
Nitric acid	HNO_3			X
Sodium carbonate	$NaCO_3$	Soda ash	X	
Sodium hydroxide	NaOH	Caustic soda	X	
Sulfuric acid	H_2SO_4			X

TABLE 12.4 Chemical Compounds Used for Alkalinity Addition

Chemical Name	Formula	Common Name	$CaCO_3$ Equivalent
Calcium carbonate	$CaCO_3$	Limestone	1.00
Calcium hydroxide	$Ca(OH)_2$	Lime	1.35
Sodium bicarbonate	$NaHCO_3$	Baking soda	0.60
Sodium carbonate	Na_2CO_3	Soda ash	0.94
Sodium hydroxide	NaOH	Caustic soda	1.25

production of sulfate (SO_4^{2-}) that represents a potential source of sulfides and hydrogen sulfide, if sulfate reduction occurs.

CHEMICALS FOR ALKALINITY ADDITION

Alkalinity addition is most often practiced in wastewater treatment systems to ensure proper nitrification. There are several chemical compounds that may be used for alkalinity addition. These compounds are listed in Table 12.4. Bicarbonate is the preferred alkalinity for nitrifying bacteria.

CHEMICALS FOR PHOSPHORUS REMOVAL

There are several chemical compounds that may be added to an SBR to remove (precipitate) orthophosphate from solution in order to satisfy a discharge requirement for total phosphorus. These chemical compounds and their impact upon pH in an SBR are listed in Table 12.5.

When chemical compounds are added to an SBR to remove phosphorus, the addition is called simultaneous precipitation of phosphorus. Simultaneous precipitation of phosphorus has the following advantage and disadvantage:

Advantage. Precipitated phosphorus is incorporated into sludge and improves its settleability.

TABLE 12.5   Chemical Compounds Used for Phosphorus Removal

Chemical Name	Formula	Common Name	pH Change
Aluminum sulfate	$Al_2(SO_4)_3 \cdot 16H_2O$	Alum	Decrease
Calcium hydroxide	$Ca(OH)_2$	Lime	Increase
Calcium oxide	CaO	Lime	Increase
Ferric chloride	$FeCl_3$	Chloride of iron	Decrease
Ferric sulfate	$Fe_2(SO_4)_3 \cdot 3H_2O$	Iron sulfate	Decrease
Sodium aluminate	$Na_2Al_2O_4$	Soda alum	Decrease

Disadvantage. Decrease in alkalinity and pH occur and may inhibit floc formation, nitrification, and enzymatic activity.

COAGULANTS

Coagulants or metal salts are added to wastewater treatment systems to capture solids, improve settleability, and dewater sludge. Commonly used coagulants consist of aluminum sulfate, ferric chloride ($FeCl_3$), and calcium hydroxide ($Ca(OH)_2$). Aluminum sulfate and ferric chloride decrease the pH of an SBR, while calcium hydroxide increases the pH of an SBR.

FLOCCULANTS

Flocculants also are added to wastewater treatment systems to capture solids, improve settleability, and dewater sludge. Cationic or positively charged polyacrylamide polymers are most commonly used. When these polymers degrade in an SBR, ammonium (NH_4^+) is released. The production of ammonium increases the alkalinity and pH of an SBR.

DISINFECTANTS

Disinfectants are added to an SBR to control undesired filamentous organism growth. Although 50% hydrogen peroxide (H_2O_2) is occasionally used, its cost and difficulty in adding to the SBR restrict its use. Chlorine is more commonly used than hydrogen peroxide. It is relatively inexpensive, easily applied, and usually available on-site as a disinfectant for the effluent.

Chlorine may be added as (1) gaseous chlorine (Cl_2) dissolved in water, (2) calcium hypochlorite ($Ca(OCl)_2$), or (3) sodium hypochlorite (Na_2OCl). The addition of dissolved chlorine produces hypochlorous acid (HOCl) and hypochlorite ions (OCl^-) that lower the pH in an SBR. Hypochlorous acid and hypochlorite ions make up the free chlorine residual.

The addition of calcium hypochlorite or high test hypochlorite (HTH) and sodium hypochlorite or bleach produces hypochlorite ions. Because an alkali compound is added to calcium hypochlorite and sodium hypochlorite to enhance their stability, the addition of these chemical compounds increases the pH of the SBR.

TABLE 12.6   Chemical Compounds Suitable for Nutrient Addition

	Chemical Compound	
Nutrient Needed	Compound	Formula
Nitrogen	Anhydrous ammonia	NH_3
	Aqua ammonia	NH_4OH
	Ammonium bicarbonate	NH_4HCO_3
	Ammonium carbonate	$(NH_4)_2CO_3$
	Ammonium chloride	NH_4Cl
	Ammonia sulfate	$(NH_4)_2SO_4$
Phosphorus	Trisodium phosphate	Na_3PO_4
	Disodium phosphate	Na_2HPO_4
	Monosodium phosphate	$Na_3H_2PO_4$
	Sodium hexametaphosphate	$Na_3(PO_4)_6$
	Sodium tripolyphosphate	$Na_5P_3O_{10}$
	Tetrasodium pyrophosphate	$Na_4P_2O_7$
	Phosphoric acid	H_3PO_4
Nitrogen and/or phosphorus	Ammonium phosphate	$NH_4H_2PO_4$

CARBON SOURCES FOR DENITRIFICATION

In order for acceptable denitrification to occur during an anoxic period (Static Fill Phase or denitrification period after React Phase), an adequate quantity of soluble cBOD or carbon must be available. If the cBOD of the wastewater does not contain an adequate quantity of carbon, then a carbon source must be added. There are several carbon sources or chemical compounds that are used for denitrification. The most commonly used carbon sources include acetate (CH_3COOH), sodium acetate ($NaOOHCCH_3$), ethanol (CH_3CH_2OH), methanol (CH_3OH), and glucose ($C_6H_{12}O_6$). Acetate and sodium acetate lower the pH of an SBR.

CHEMICALS FOR NUTRIENT ADDITION

Nutrient deficiencies usually occur in an SBR for nitrogen and phosphorus when the SBR receives a relatively large quantity of industrial wastewater that is rich in soluble cBOD but deficient for nitrogen and/or phosphorus. Nutrient deficiencies often are corrected by the addition of chemical compounds that release ammonium for nitrogen or orthophosphate for phosphorus. Chemicals used for addition of ammonium and orthophosphate are listed in Table 12.6.

Due to the relatively small quantity of chemical compounds that are added to an SBR to correct for a nutrient deficiency, nutrient addition results in relatively small changes in alkalinity and pH. However, overdosing of a chemical compound or spills of a chemical compound may result in significant changes in alkalinity and pH.

CHEMICALS DISCHARGED TO AN SBR

Small-capacity SBRs are vulnerable to significant changes in alkalinity and pH from relatively small quantities of specific chemical compounds discharged to an SBR. This is especially true during low loading conditions.

Specific chemical compounds that may be accidentally discharged or discharged in unacceptable quantities include cleaning agents, disinfectants, lipids (fats, oils, and grease) and solvents. These compounds or their degradation may result in significant changes in the alkalinity and pH of an SBR. Therefore, an inspection of all industrial and commercial dischargers should be performed to determine the chemical compounds that are or may be discharged. Also, the collection and review of all material safety data sheets (MSDS) of chemicals present or used at the commercial or industrial establishment should be performed. The operator of the SBR should keep the MSDS on file for review and update.

TROUBLESHOOTING KEY FOR THE IDENTIFICATION OF OPERATIONAL FACTORS RESPONSIBLE FOR UNDESIRED CHANGES IN PH AND ALKALINITY

1. Has a slug discharge or excessive quantity of soluble cBOD occurred resulting in the release of a relatively large quantity of carbon dioxide and the formation of carbonic acid that lowers the pH of the SBR?

   Yes   See 2
   No   See 3

2. Terminate or equalize the discharge of soluble cBOD and add an alkali compound to increase the pH to 6.8–7.2.

3. Has unexpected nitrification occurred in the SBR or the rate of nitrification increased?

   Yes   See 4
   No   See 5

4. Collect a sample of mixed liquor immediately after the React Phase, filter the solids through a Whatman® No. 4 filter paper and test the filtrate for ammonium, nitrite, and nitrate. If nitrite or nitrate is detected at ≥1 mg/L, the SBR is nitrifying. If nitrification is not desired, then terminate nitrification, for example, reduce aeration or MLSS concentration. If ammonium concentration is less than that required by the SBR discharge permit, then reduce nitrification by decreasing dissolved oxygen concentration during the React Phase or add appropriate chemicals for pH and alkalinity adjustment. However, it is necessary to monitor and adjust operational conditions to maintain acceptable nitrification (permit compliance).

5. Has sulfide or fermentative wastes entered the SBR, or has sulfate reduction and fermentation occurred in the SBR?

   Yes   See 6
   No   See 7

6. Check the influent or sewer system for atypical quantities of sulfate, sulfide, or septic conditions such as accumulated solids in sewers or lift stations. Terminate

the source of high quantities of sulfate and sulfides and clean and correct the factors responsible for the septic condition. If sulfate reduction or fermentation occurred in the SBR, check phase times to ensure that dissolved oxygen and nitrates were not exhausted. Decrease retention time, increase dissolved oxygen and/or nitrate concentration, and adjust the pH of the SBR. The use of a redox (ORP) probe may be helpful in locating a section of the sewer system, tank, or recycle stream that has a septic condition—that is, a redox $\leq -100 \, mV$.

7. Has sulfide oxidation occurred?

   Yes   See 8
   No   See 9

8. If atypical concentrations of sulfide are entering the SBR, identify and correct the source of sulfide discharge. Sulfide may be chemically oxidized to elemental sulfur by chlorination before it enters the SBR.

9. Are chemical compounds being added to the SBR for process control—for example, pH control, alkalinity addition, phosphorus removal, coagulation, flocculation, disinfection, carbon addition, or nutrient addition?

   Yes   See 10
   No   See 11

10. Ensure that the preparation of the chemical compound and calculated chemical feed dose are correct. Also, ensure that the feed rate of the chemical compound is properly adjusted for changes in loading conditions. Check all equipment and lines to prevent leaks and spills of the chemical compound.

11. Check the influent to the treatment facility for atypical odors, solids, foam, pH, and ORP values. The use of conductivity readings also may indicate an atypical condition in the influent. In order to use pH, ORP, and conductivity as indicators of atypical influent, sampling and monitoring of the influent should be performed on a routine schedule to establish a baseline of values for each parameter. If the influent is atypical, determine the source of discharge of the waste responsible for the condition in the influent. The discharge should be terminated or regulated.

Check List for the Identification of Operational Factors Responsible for Undesired Changes in pH and Alkalinity

Operational Factor	√ If Monitored	√ If Possible Factor
Respiratory release of CO_2 during slug discharge of soluble cBOD		
Nitrification		
Sulfate reduction		
Sulfide oxidation		
Fermentation		
Chemical compounds added to the SBR		
Chemical compounds discharged to the SBR		

13

Troubleshooting Foam and Scum Production

BACKGROUND

The production of foam in a sequencing batch reactor is the result of undesired bacterial activity or chemical discharge. Foam consists of entrapped air or gas bubbles beneath a thin layer of solids. Typical gases entrapped in foam consist of those produced through bacterial degradation of cBOD and consist of carbon dioxide (CO_2), molecular nitrogen (N_2), and nitrous oxide (N_2O).

There are several operational conditions that are responsible for the production of foam in an SBR. These conditions include:

- Erratic sludge wasting rates
- Undesired growth of foam-producing filamentous organisms
- Nutrient deficiency
- Sludge aging, including young sludge age
- Slug discharge of soluble cBOD
- Excess surfactants
- Excess polymer
- Viscous floc or Zoogloeal growth
- Excess fine solids

Foam in an SBR may be described by its texture and color (Table 13.1). The texture and color of foam may differ from its description in the table due to the

Troubleshooting the Sequencing Batch Reactor, by Michael H. Gerardi
Copyright © 2010 by John Wiley & Sons, Inc.

TABLE 13.1 Foam Production in an SBR

Contributing Operational Condition	Texture and Color
Erratic sludge wasting rates	Concentric circles of viscous dark brown and viscous light brown foam appear after termination of aeration and mixing
Foam-producing filamentous organisms	Viscous chocolate-brown
Nutrient deficiency	Greasy gray (old sludge) or billowy white (young sludge)
Sludge aging	With increasing sludge age: billowy white to crisp white to crisp brown to viscous dark brown
Slug discharge of soluble cBOD	Billowy white
Excess surfactants	Billowy white
Excess polymer	Billowy white
Excess fine solids	Pumic-like gray

sludge age of the SBR. With increasing sludge age, foam becomes more viscous and darker in color; with decreasing sludge age, foam becomes more billowy and lighter in color.

When air and gas bubbles escape from foam, foam collapses. The collapsed foam often is referred to as scum. Collapsed foam should not be confused with the development of brown "flakey" scum on the surface of the decant. Here, brown flakey scum or soap may appear after the die-off of large numbers of bacteria. There is no entrapped air or gas bubbles beneath the scum.

When bacteria die, they undergo autolysis; that is, they break open. Due to autolysis, much of the intracellular components of the bacteria are released to the bulk solution. Some of the components are fatty acids. Fatty acids are short—chain organic compounds such as formate (CH_3COOH), acetate (CH_3CHOOH), and propionate (CH_3CH_2COOH). When fatty acids combine with calcium ions (Ca^{2+}) in the bulk solution, an insoluble soap is produced. Domestic wastewater typically has approximately 150 mg/L of soluble calcium ions.

Bacteria die in large numbers for two reasons. First, they starve when adequate substrate is not present. Second, they experience toxicity.

ERRATIC SLUDGE WASTING RATES

Frequent removal or wasting of highly variable quantities of solids or mixed liquor suspended solids (MLSS) from an SBR over several weeks may result in the production of "pockets" of young bacterial growth and old bacterial growth. Because young bacteria do not produce large quantities of oils as compared to old bacteria, air and gas bubbles captured by young and old solids result in the production of different foam for each bacterial population. Young bacteria produce viscous light brown foam, whereas old bacteria produce viscous dark brown foam. Concentric circles of light brown foam and dark brown foam can be observed in an SBR when aeration and mixing are terminated.

FOAM-PRODUCING FILAMENTOUS ORGANISMS

There are three foam-producing filamentous organisms. These organisms are *Microthrix parvicella*, Nocardioforms, and type 1863. Foam typical of these organisms is viscous chocolate-brown.

Foam production by filamentous organisms can occur by the secretion of hydrophobic compounds or the production of lipids and biosurfactants. For example, *Microthrix parvicella* produces hydrophobic compounds that capture air and gas bubbles that result in foam production. Live Nocardioforms produce lipids that coat floc particles and capture air and gas bubbles, while dead Nocardioforms undergo autolysis and release biosurfactants that alter the surface tension of the wastewater and permit foam production.

NUTRIENT DEFICIENCY

Nutrient deficiencies usually occur when the SBR receives relatively large quantities of industrial wastewaters that are rich in soluble cBOD (Table 13.2) but lack adequate quantities of nitrogen and phosphorus.

During a nutrient deficiency, bacteria in floc particles absorb soluble cBOD. However, due to the lack of nitrogen or phosphorus, the soluble cBOD cannot be degraded, and the cBOD is converted to an insoluble polysaccharide and stored outside the bacteria. When nutrients become available, the stored polysaccharides are solubilized and degraded. Until the polysaccharides are solubilized, they capture air and gas bubbles, resulting in the production of foam upon mixing.

TABLE 13.2 Nutrient Deficient Industrial Wastewaters

Industrial Wastewater	Nitrogen-Deficient	Phosphorus-Deficient
Bakery	X	
Beverage, malt	X	X
Beverage, distilled spirits	X	X
Beverage, wine	X	X
Beverage, soda drink	X	X
Citrus		X
Coffee	X	X
Coke ovens		X
Corn	X	
Cotton kerning	X	
Dairy, milk		X
Dairy, cottage cheese	X	
Food processing	X	X
Formaldehyde	X	X
Fruit and vegetable	X	X
Leather tanning		X
Petroleum		X
Pharmaceutical		X
Phenols	X	
Pulp and paper	X	X
Textile	X	

TABLE 13.3   Chemical Compounds Suitable for Nutrient Addition

	Chemical Compound	
Nutrient Needed	Common Name	Formula
Nitrogen	Anhydrous ammonia	NH_3
	Aqua ammonia	NH_4OH
	Ammonium bicarbonate	NH_4HCO_3
	Ammonium carbonate	$(NH_4)_2CO_3$
	Ammonium chloride	NH_4Cl
	Ammonia sulfate	$(NH_4)_2SO_4$
Phosphorus	Trisodium phosphate	Na_3PO_4
	Disodium phosphate	Na_2HPO_4
	Monosodium phosphate	NaH_2PO_4
	Sodium hexametaphosphate	$Na_3(PO)_6$
	Sodium tripolyphosphate	$Na_5P_3O_{10}$
	Tetrasodium pyrophosphate	$Na_4P_2O_7$
	Phosphoric acid	H_3PO_4
Nitrogen and/or phosphorus	Ammonium phosphate	$NH_4H_2PO_4$

Different nutrient-deficient foam with respect to texture and color is produced at different sludge ages. Billowy white foam is produced at a young sludge age, while greasy gray foam is produced at an old sludge age. The difference in texture and color is due to the accumulation of oils in floc particles that are transferred to foam. Young bacteria produce a relatively small quantity of oils that accumulate in floc particles, while old bacteria produce a relatively large quantity of oils. Therefore, foam produced from floc particles at an old sludge age is more viscous and darker in color than foam produced from floc particles at a young sludge age.

Nutrient deficiencies often are corrected by the addition of chemical compounds that release ammonium for a nitrogen deficiency and orthophosphate for a phosphorus deficiency. Chemicals available for use to correct a nutrient deficiency for nitrogen and phosphorus are listed in Table 13.3.

SLUDGE AGING

There are several types of foam that are produced through sludge aging. The quantity of foam produced may or may not be problematic and may be controlled by increasing or decreasing MLVSS. The sequence of foam production from start-up, aging, and maturation of the MLVSS typically is as follows: billowy white to crisp white to crisp brown to viscous dark brown.

Billowy white foam is produced at a young sludge age when the bacterial population is relatively small—for example, <1000 mg/L MLVSS. At a relatively small population size, the bacteria are not able to adequately degrade the surfactants entering the SBR. Therefore, billowy white foam appears. Operational conditions that mimic a young sludge and permit the development of billowy white foam from surfactants include (1) inhibition or toxicity, (2) recovery from inhibition or toxicity, and (3) over-wasting of solids (MLVSS). When an adequate concentration of MLVSS (>1000 mg/L) is present, the bacteria in the mixed liquor are capable of degrading the surfactants and crisp white foam appears.

As the bacterial population ages, more oils are produced by the bacteria. These oils accumulate in the floc particles and are transferred to foam. Therefore, the foam becomes crisp brown. As the sludge age continues to increase, foam-producing filamentous organisms may proliferate. The lipids and hydrophobic compounds released by these organisms and the oils secreted by old bacteria result in the production of viscous chocolate-brown foam.

SLUG DISCHARGE OF SOLUBLE cBOD

A slug discharge of soluble cBOD is considered to be a quantity of cBOD that is two to three times greater than the normal quantity of cBOD that is received over a 2- to 4-hr period. Slug discharges usually are associated with industrial wastewaters, but the rapid transfer or discharge of septic wastewater to an SBR can mimic a slug discharge.

In the presence of adequate nutrients and dissolved oxygen, bacteria rapidly degrade the soluble cBOD. This results in young bacterial growth and the production of a copious quantity of insoluble polysaccharides that capture air and gas bubbles. Because young bacterial growth has only a relatively small quantity of oils that are incorporated into floc particles, the foam produced from a slug discharge of soluble cBOD is billowy white.

EXCESS SURFACTANTS

The presence of excess surfactants in an SBR results in a change in the surface tension of the wastewater that permits foam production. Surfactant foam is billowy white. Excess surfactants produce foam, because an appropriate number of bacteria are not present to degrade the surfactants. Operational conditions that permit the development of surfactant foam include (1) low MLVSS, (2) toxicity, (3) recovery from toxicity, and (4) over-wasting of MLVSS.

In addition to foam production, surfactants represent several additional operational concerns. These concerns include:

- Dispersion of floc particles and loss of fine solids
- Inhibition and toxicity
- Inefficient oxygen transfer in fine aeration systems

Many surfactants such as the sulfur-containing or sulfonated detergents represent an additional operational concern, namely, the release of sulfate (SO_4^{2-}) when the detergents are degraded in an SBR (Figure 13.1).

EXCESS POLYMER

Cationic polymers are commonly used at activated sludge processes for sludge dewatering, sludge thickening, and solids capture. Some polymers, especially the

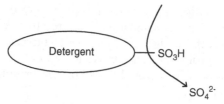

Figure 13.1 *Sulfonated detergent. Sulfonated detergents contain sulfur in a key functional group, HSO₃. When sulfonated detergents are degraded, this group is released to the bulk solution and forms sulfate, SO₄²⁻.*

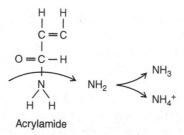

Figure 13.2 *Release of ammonium from polyacrylamide polymer. Polyacrylamide polymers contain acrylamide (H₂NCOHCHCH). When acrylamide is degraded, the amino group (-NH₂) is released and forms ammonia (NH₃) at a high pH (≥9.4) or forms ammonium (NH₄⁺) at a low pH.*

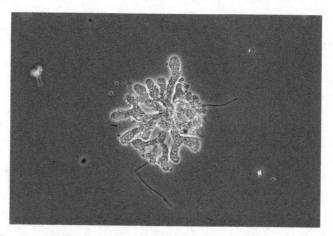

Figure 13.3 *Dendritic Zoogloeal growth. Dendritic ("tooth-like" or "finger-like") Zoogloeal growth is less often observed in activated sludge processes then amorphous Zoogloeal growth. Dendritic usually appears at young sludge ages or mean cell residence times (MCRT).*

polyacrylamides polymers, contain amino groups (-NH$_2$) that release ammonium (NH$_4^+$) when the polymers are degraded (Figure 13.2). Ammonium represents an increase in alkalinity and change in the surface tension of the wastewater.

VISCOUS FLOC

Viscous floc or Zoogloeal growth is the rapid and undesired proliferation of floc-forming bacteria. This growth may appear in the dendritic form (Figure 13.3) or the

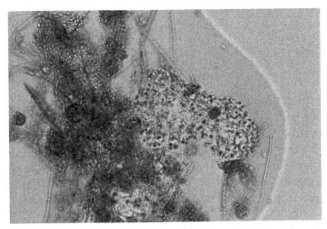

Figure 13.4 *Amorphous Zoogloeal growth. Under a Gram-stained smear of amorphous Zoogloeal growth, the loosely compacted floc-forming bacteria can be observed as extending from the right side of the floc particle. The Gram stain reveals the presence of the cells and the relatively large quantity of gelatinous material that surrounds and separates the bacterial cells.*

amorphous form (Figure 13.4). The development of viscous floc results in the production of a copious quantity of gelatinous material that captures air and gas bubbles. This permits the production of billowy white foam. The presence of undesired viscous floc or Zoogloeal organisms is associated with (1) a septic condition upstream of the SBR or the use of Static Fill Phase for biological phosphorus release, (2) high MCRT, (3) nutrient deficiency, (4) organic acids and (5) high F/M.

EXCESS FINE SOLIDS

The presence of excess fine solids during the React Phase may produce pumic-like gray foam. The fine solids that enter the SBR usually are from the discharge of other treatment processes such as the anaerobic digester or dewatering operations where poor solids capture occurs.

TROUBLESHOOTING KEY FOR IDENTIFICATION OF OPERATIONAL FACTORS RESPONSIBLE FOR FOAM AND SCUM PRODUCTION

1. Has sludge been wasted in an erratic pattern over a relatively short period of time?

   Yes   See 2
   No   See 3

2. Remove small quantities of sludge over as long a period of time as possible to stabilize the age of the bacterial population.

3. Does foam contain a relatively large population of foam-producing filamentous organisms?*

   Yes   See 4
   No   See 5

4. Filamentous organism growth should be controlled in foam and mixed liquor.† Foam may be sprayed with a 10% to 15% sodium hypochlorite solution. Allow the solution to remain in contact with the foam for 1–2 hr and then collapse the foam with a water spray. Foam may be controlled with the addition of a poly-glycol defoaming agent or other appropriate defoaming agent. Foam produced by Nocardioforms also may be treated with bioaugmentation products that produce lipase, the enzyme for degrading fats, oils, and grease.

5. Does a nutrient deficiency exist for nitrogen and/or phosphorus in the mixed liquor?‡

   Yes   See 6
   No   See 7

6. Foam may be treated with a water spray or defoaming agent. The source of the nutrient deficiency should be identified and regulated, and appropriate nutri-ents added to either the nutrient-deficient wastewater or the SBR.

7. Is foam a result of sludge aging?

   Yes   See 8
   No   See 9

8. Increasing or decreasing MCRT may correct foam produced through sludge aging. Increasing the sludge age, is achieved by reducing or terminating the quantity of solids (MLSS) removed from the SBR, while decreasing the sludge

---

*The presence of foam-producing filamentous organisms can be detected by examining a Gram-stained smear of foam under the microscope. The two most commonly observed foam-producing filamentous organisms are Nocardioforms and *Microthrix parvicella*. Nocardioforms are Gram positive (blue) and highly branched, while *Microthrix parvicella* is Gram positive (blue) and nonbranched and has a blue "bead-like" appearance along the length of the filament.

†Control of Nocardioforms in the SBR can be achieved through several measures, including the use of the Static Fill Phase and the Mix Fill Phase.

‡To determine the presence of an adequate quantity of nitrogen or phosphorus in an SBR, a grab sample of mixed liquor should be taken immediately after the React Phase. The sample should be filtered through a Whatman® No. 4 filter paper. The filtrate should be tested for ammonical-nitrogen (NH_4^+-N), nitrate-nitrogen (NO_3^--N), and orthophosphate-phosphorus ($H_2PO_4^-$/HPO_4^{2-}). If the concentration of ammonical-nitrogen is ≥1 mg/L or the concentration of nitrate-nitrogen is ≥3 mg/L, a nutrient deficiency for nitrogen does not exist. If the concentration of orthophosphate-phosphorus is ≥0.5 mg/L, a nutrient deficiency for phosphorus does not exist. However, an acceptable, specific oxygen uptake rate (SOUR) should be performed to determine if a toxic condition exists in the SBR that would permit elevated concentrations of ammonical-nitrogen and orthophosphate-phosphorus in the filtrate.

age, is achieved by increasing the quantity of solids (MLSS) removed from the SBR.

9. Has a slug discharge of soluble cBOD occurred?

Yes   See 10
No   See 11

10. Identify and regulate the slug discharge of soluble cBOD. The billowy white foam produced from the slug discharge may be collapsed with a water spray.

11. Has a discharge of excess surfactants occurred?

Yes   See 12
No   See 13

12. Identify and regulate the discharge of excess surfactants. Foam produced from surfactants may be controlled with the use of bioaugmentation products that release surfactant-degrading enzymes.

13. Has a discharge of cationic polyacrylamides polymer occurred?

Yes   See 14
No   See 15

14. Identify and regulate the discharge of the polymer. Foam production from polymer may be controlled with the use of a water spray.

15. Is foam production due to viscous floc or Zoogloeal growth?

Yes   See 16
No   See 17

16. Billowy white foam may occur due to the growth of Zoogloeal organism or viscous floc.* Most Zoogloeal organisms are strict aerobes, and their growth can be controlled with the use of Mix Fill Phase. Foam may be collapsed with a water spray.

17. Scum production may be due to collapsed foam. Identify foam in the SBR and correct for foam production accordingly. Brown flaky scum is produced through the die-off of large numbers of bacteria due to high MLVSS or toxicity. Either reduce solids (MLSS) inventory, if scum production is due to a decrease in cBOD loading, or identify and terminate the source of the toxic discharge to the SBR.

---

*Viscous floc or Zoogloeal growth can be identified by a microscopic examination. Viscous floc often can be observed on the walls of the SBR. Here, a slimy white or grayish-white growth may be found. The growth may have a black "peppering" effect.

Check List for the Identification of Operational Factors Responsible for Foam and Scum Production

Operational Factor	√ If Monitored	√ If Possible Factor
Erratic sludge wasting rate		
Foam-producing filamentous organisms		
Nutrient deficiency		
Sludge aging		
Slug discharge of soluble cBOD		
Excess surfactants		
Excess polymer		
Viscous floc		

14

Troubleshooting Low Dissolved Oxygen

BACKGROUND

Dispersed or suspended bacteria that utilize free molecular oxygen (O_2) function properly at a dissolved oxygen (DO) concentration of <0.3 mg/L. However, when flocculated, the concentration of DO outside the floc particle must be able to penetrate to the core of the floc particle and provide an adequate dissolved oxygen concentration in the core.

Dissolved oxygen is supplied to any activated sludge process, including a sequencing batch reactor to satisfy six basic biological wastewater treatment needs. These needs are:

- Promote floc formation
- Ensure endogenous respiration
- Oxidize (degrade) cBOD
- Oxidize (degrade) nBOD
- Prevent the undesired growth of low DO filamentous organisms
- Achieve biological phosphorus uptake

PROMOTE FLOC FORMATION

Floc formation is the development of dense, firm, and mature floc particles or solids (Figure 14.1). This condition provides for (1) desired settleability and compaction of solids during the Settle Phase and (2) resistance to shearing action (excess

Troubleshooting the Sequencing Batch Reactor, by Michael H. Gerardi
Copyright © 2010 by John Wiley & Sons, Inc.

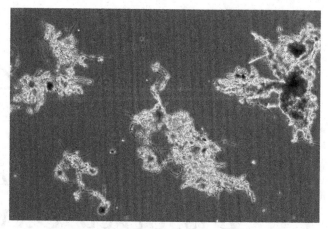

Figure 14.1 *Mature floc particle. A mature floc particle typically is medium (150–500 μm) or large (≥500 μm) in size and irregular in shape due to the growth of a limited number (1 to 5) filamentous organisms. The floc particle also is golden-brown in the core due to the accumulation of oils from old bacteria cells and light in color at the perimeter due to lack of production of oils from young bacterial cells. The floc particle is firm or dark blue under methylene blue staining, and it tests negatively to the India ink reverse stain. There is no significant Zoogloeal growth, interfloc bridging, or open floc formation. The bulk solution around the floc particle is clean; that is, there is little dispersed growth and particulate material. The floc particle may have crawling and/or stalk ciliates on its surface.*

turbulence) and loss of fine solids produced through aeration and mixing of solids. The solids are referred to as mixed-liquor suspended solids (MLSS) and consist of nonliving and living components. Most of the living components are bacteria. Because bacteria are volatile, they are referred to as mixed-liquor volatile suspended solids (MLVSS) and represent approximately 65% to 75% of the MLSS.

A typical, mature floc particle is usually >150 μm in size, irregular in shape, and golden-brown in color. The floc particle contains some filamentous organisms (approximately 1–5) that are found either in the particle or extending into the bulk solution from the perimeter of the particle.

If an SBR experiences no adverse operational condition, then floc formation can be achieved at a dissolved oxygen concentration of ≥1 mg/L. Initiation of floc formation with domestic or municipal wastewater takes approximately 1 day of MCRT, while initiation of floc formation with industrial wastewater takes approximately 3 days of MCRT. Maturation of floc particles takes approximately 7–10 days of MCRT.

ENSURE ENDOGENOUS RESPIRATION

Endogenous respiration is the bacterial degradation (oxidation) of stored food. Many bacteria store food as (1) a slime layer that coats the cell or (2) starch granules that are deposited inside and outside the cell (Figure 14.2). Stored food is degraded in an SBR during any of the following conditions:

- After degradation of influent soluble cBOD
- Significant reduction of influent soluble cBOD
- Significant increase in the quantity of colloidal or particulate cBOD accompanied by a decrease in the quantity of soluble cBOD
- During biological phosphorus release

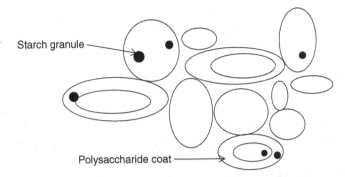

Figure 14.2 *Food stored by bacterial cells. There are two major means of food storage that are used by bacterial cells. These consist of the production of starch granules and a polysaccharide coat. The starch granules may be found inside the cell or inside the polysaccharide coat. Not all bacteria, especially filamentous bacteria, are capable of storing food.*

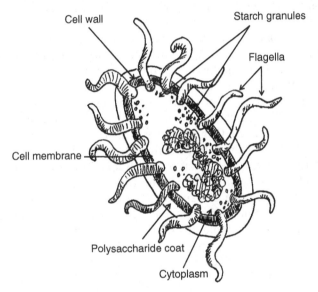

Figure 14.3 *Digestion of cytoplasm, basal respiration. After bacterial cells have degraded starch granules and the polysaccharide coat during endogenous respiration, endogenous respiration may continue if the bacterial cell starts to digest itself; that is, it digests its cytoplasm or "gut" content. When this occurs, the continuation of endogenous respiration is referred to as basal respiration.*

Endogenous respiration results in the consumption of DO. With increasing MLVSS inventory, a larger number of bacteria and a larger quantity of stored food are present in the SBR. Therefore, endogenous respiration consumes large quantities of DO with increasing MLVSS. Proper degradation of stored food under endogenous respiration occurs at a DO value of $\geq 0.8\,mg/L$.

When stored food has been exhausted, bacteria begin to degrade (digest) their cytoplasm (Figure 14.3). This is known as basal respiration and is a continuation of endogenous respiration. Basal respiration also consumes large quantities of DO. Endogenous respiration and the occurrence of basal respiration significantly decrease the volatile content of the MLSS. The decrease in volatile content promotes better settling of solids. Endogenous respiration and basal respiration are promoted in a sequencing batch reactor with increasing MCRT and MLVSS and decreasing food-to-microorganism (F/M) ratio.

OXIDIZE (DEGRADE) OF cBOD

Oxidation or degradation of cBOD may occur rapidly or slowly in an SBR. The rate of degradation is dependent upon several operational conditions, including the types of cBOD present and dissolved oxygen concentration. Carbonaceous BOD consists of organic compounds—that is, compounds that contain carbon and hydrogen. Examples of soluble cBOD include methanol (CH_3OH), ethanol (CH_3CH_2OH), acetate (CH_3COOH), and sucrose ($C_6H_{12}O_6$). These compounds represent alcohols (methanol and ethanol), acids (acetate), and carbohydrates (sucrose).

There are several types of cBOD. These include soluble forms that degrade quickly (<7 hr), colloidal and particulate BOD, and lipids that degrade slowly. Colloids are large, insoluble complex compounds such as proteins. Particulates are also large, insoluble complex compounds such as cellulose or starch. In order for colloidal and particulate cBOD to be degraded in an SBR, they must first be adsorbed to the bacteria and then solubilized to simplistic compounds that are absorbed by bacteria. Degradation of cBOD occurs inside bacterial cells.

Lipids consist of fats, oils, and grease that are not soluble in wastewater. Lipids that are simple in structure may be absorbed by bacteria and degraded slowly, while lipids that are complex in structure must be adsorbed by bacteria, solubilized, absorbed, and then degraded. Generally, lipids degrade more slowly than colloidal and particulate BOD, and these three forms of cBOD degrade more slowly than soluble cBOD. Therefore, the degradation of cBOD is most rapid when the quantity and quality of cBOD is mostly soluble. The rapid degradation of soluble cBOD exerts an immediate and large demand for dissolved oxygen and nutrients.

Approximately 1.8 pounds of oxygen are consumed for each pound of cBOD degraded. Degradation of soluble cBOD takes approximately 0.3 day of MCRT. Degradation of colloidal and particulate cBOD takes approximately 2 days of MCRT, while the degradation of complex lipids may take more than 4 days of MCRT.

OXIDIZE (DEGRADE) nBOD

Oxidation or degradation of nBOD is nitrification. It is the biological oxidation of ammonium (NH_4^+) to nitrite (NO_2^-) and then to nitrate (NO_3^-). After adequate degradation of cBOD, nitrification may occur. The commonly accepted concentration for cBOD in an activated sludge process that permits nitrification is <40 mg/L.

The rate of nitrification is very rapid at dissolved oxygen values of ≥2 mg/L. With increasing dissolved oxygen concentration, nitrification accelerates. The maximum rate of nitrification occurs at approximately 3 mg/L. However, nitrification may improve at dissolved oxygen values of >3 mg/L, if the higher dissolved oxygen concentration permits more rapid removal of cBOD during the React Phase. This would provide more time for nitrification to occur. Approximately 4.6 pounds of oxygen are consumed for each pound of nBOD as ammonium that is degraded to nitrate.

Nitrogenous BOD consists of ammonium (NH_4^+) and nitrite (NO_2^-). Ammonium and ammonia are reduced forms of nitrogen—that is, nitrogen that is bonded to hydrogen. Ammonium is present in wastewater in large quantities at pH values of <9.4, while ammonia is present in wastewater in large quantities at pH values of ≥9.4. When testing and reporting the concentration of reduced nitrogen in wastewater, the analytical technique requires that the pH of the wastewater sample be increased with an alkali compound to a value of >9.4. The addition of an alkali compound converts ammonium to ammonia that is either measured by an immersed ammonia probe or stripped to the atmosphere by mixing action and measured by an ammonia probe positioned over the wastewater. By testing for ammonia, the concentration of ammonium is indirectly measured.

Domestic wastewater typically has approximately 30 mg/L of ammonium. However, a very large quantity of ammonium is released in the SBR when organic-nitrogen compounds are degraded. These compounds contain an amino group ($-NH_2$) that is released during bacterial degradation of the organic-nitrogen compound. The released amino group forms mostly ammonium in the SBR at pH values of <9.4. The quantity of ammonium that is capable of being produced from organic-nitrogen compounds can be determined by the total Kjeldahl nitrogen (TKN) test and an ammonia test.

The TKN test measures the quantity of ammonium that is in the wastewater and the quantity of ammonium that can be released from organic-nitrogen compounds. Therefore, the organic-nitrogen value of a wastewater sample is the TKN minus the value for ammonium. The total nitrogen value of a wastewater sample is the TKN, nitrite, and nitrate (Figure 14.4).

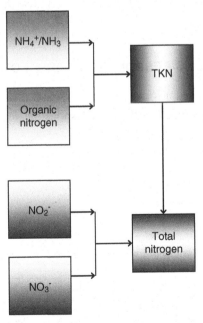

Figure 14.4 *Total nitrogen in wastewater. Total nitrogen in wastewater consists of total Kjeldahl nitrogen (TKN), nitrite (NO_2^-), and nitrate (NO_3^-). TKN consists of ammonium (NH_4^+)/ammonia (NH_3) and organic nitrogen.*

Nitrite does exert an oxygen demand. It is oxidized to nitrate. Nitrate does not exert an oxygen demand. Nitrate is completely oxidized. Nitrite may enter an SBR from an industrial discharge.

PREVENT THE UNDESIRED GROWTH OF LOW DO FILAMENTOUS ORGANISMS

There are four filamentous organisms that proliferate under a low DO concentration. These organisms are *Haliscomenobacter hydrossis*, *Microthrix parvicella*, *Sphaerotilus natans*, and type 1701. The rapid growth of these organisms is undesired, because they contribute to settleability problems. *Microthrix parvicella* also produces problematic, viscous chocolate-brown foam.

To prevent the undesired growth of these filamentous organisms, it is necessary to maintain an adequate DO in the core of the floc particles. Therefore, as more soluble cBOD is degraded in an SBR, more DO must be provided. However, the use of a Mix Fill Phase (anoxic period) can control of some of these filamentous organisms (Table 14.1).

Because domestic wastewater contains a relatively large quantity of lipids and colloidal and particulate BOD that degrade slowly and do not exert a significant DO demand, the degradation of domestic wastewater usually does not result in the undesired growth of low DO filamentous organisms.

However, most industrial wastewaters do contain relatively large quantities of soluble cBOD that exert an immediate and large demand for DO. Therefore, SBR that receive industrial wastewater need an adequate concentration of DO at all times during React Phase to prevent the undesired growth of low DO filamentous organisms.

OPERATIONAL FACTORS RESPONSIBLE FOR A LOW DO

In addition to equipment malfunction and inadequate aeration equipment, there are five major operational factors that are responsible for a low DO in an SBR. These factors are:

- Oxygen scavengers
- Increase in organic loading
- Endogenous respiration

Table 14.1 Control of Low DO Filamentous Organisms with Mix Fill Phase

Filamentous Organism	Controlled by Mix Fill Phase
Haliscomenobacter hydrossis	Yes
Microthrix parvicella	No
Sphaerotilus natans	Yes
Type 1701	Yes

• Undesired nitrification
• Inadequate oxygen transfer

OXYGEN SCAVENGERS

Oxygen scavengers react quickly with dissolved oxygen. Often, they react with dissolved oxygen before bacteria can use dissolved oxygen for the degradation of BOD. There are two major oxygen scavengers of concern to SBR. These scavengers are sulfide (HS^-) and sulfite (SO_3^{2-}).

Corrosion inhibitors are used in boilers and cooling towers and may be found in industrial wastewaters. A commonly used corrosion inhibitor is sodium sulfite (Na_2SO_3). In boilers and cooling towers the corrosion inhibitors react with dissolved oxygen before the dissolved oxygen can react with a metal surface and cause corrosion.

Sulfide enters an SBR from an anaerobic or septic condition through two events. First, thiol groups (-SH) are released from amino acids and proteins in fecal waste when they are degraded by bacteria in biofilm and sediment in sewers, manholes, and lift stations under an anaerobic condition (Figure 14.5). The released thiol group forms sulfide (HS^-). Second, under an anaerobic condition, sulfate-reducing bacteria use sulfate (SO_4^{2-}) to degrade soluble cBOD [Eq. (14.1)]. The use of sulfate results in the production of sulfide. Sulfate enters sanitary sewers in urine and groundwater.

$$cBOD + SO_4^{2-} \xrightarrow{\text{sulfate-reducing bacteria}} CO_2 + H_2O + cells + H_2S/HS^- + OH^-$$
$$(14.1)$$

In an SBR sulfide may be biological or chemically [Eq. (14.2)] oxidized to sulfate. These oxidations result in the consumption of dissolved oxygen. The sulfur

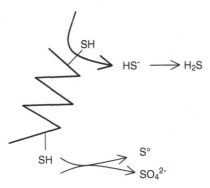

Figure 14.5 *Release of thiol group. Thiol groups (-SH) are found on the amino acid cysteine and proteins that contain the amino acid. When the amino acid and proteins are degraded, the thiol group is released and forms either sulfide (HS^-) or hydrogen sulfide (H_2S) under an anaerobic condition. However, these reduced forms of sulfur (HS^- and H_2S) can be oxidized biologically by sulfur-oxidizing bacteria or chemically to sulfate (SO_4^{2-}) and can be oxidized by sulfide-loving filamentous organisms to elemental sulfur (S^0). Sulfide-loving filamentous organisms include* Beggiatoa *spp.,* Thiothrix *spp., and type 021N.*

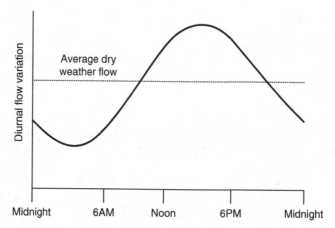

Figure 14.6 *Diurnal flow of wastewater. Wastewater flow to an SBR varies greatly through the day for domestic and municipal wastewater treatment plants. This is due to the diurnal flow pattern of residential, commercial, and industrial discharges to the sewer system. Peak hydraulic loading usually occurs between 10 AM and 6 PM, while lowest hydraulic loading usually occurs between 10 PM and 6 AM.*

filamentous organisms, *Beggiatoa* (Figure 14.6), *Thiothrix*, and type 021N, remove sulfide and oxidize it to elemental sulfur, S^0, to obtain energy. This permits their rapid proliferation. Their undesired growth contributes to settleability problems and bulking.

$$2HS^- + 4O_2 \rightarrow 2SO_4^{2-} + 2H^+ \tag{14.2}$$

INCREASE IN ORGANIC LOADING

An increase in cBOD loading, especially a slug discharge of soluble cBOD, exerts a significant increase in dissolved oxygen demand. A slug discharge is considered to be a quantity of cBOD that is two to three times greater than the typical quantity of cBOD that is received over a 2- to 4-hr period.

UNDESIRED NITRIFICATION

Undesired nitrification is the occurrence of nitrification in an SBR that is not required to nitrify—that is, satisfy an ammonia discharge limit or total nitrogen discharge limit.

INADEQUATE OXYGEN TRANSFER

The inability to transfer an adequate quantity of dissolved oxygen to bacteria in floc particles may be due to (1) changes in diurnal oxygen demand and (2) an undersized aerator. Diurnal oxygen demand is the increase and decrease in dissolved oxygen requirement in an SBR to degrade cBOD and nBOD due to variations throughout the day in wastewater flow and strength (Figure 14.6).

TROUBLESHOOTING KEY FOR THE IDENTIFICATION OF OPERATIONAL FACTORS RESPONSIBLE FOR LOW DISSOLVED OXYGEN

1. Is the aeration equipment malfunctioning?

   Yes   See 2
   No   See 3

2. Inspect and make necessary repairs or maintenance.

3. Does the influent contain oxygen scavengers?

   Yes   See 4
   No   See 5

4. If the influent contains corrosion inhibitors, these inhibitors should not be discharged to the SBR. If the influent contains sulfides, identify and correct the source of sulfides or chlorinate the influent. Chlorination of sulfide oxidizes the sulfide to elemental sulfur (S^0).

5. Is there an increase in cBOD loading or slug discharge of soluble cBOD?

   Yes   See 6
   No   See 7

6. Increased cBOD loading should be equalized as much as possible over 24 hr. This may be achieved with an equalization tank. If a slug discharge of soluble cBOD is responsible for the low DO, the slug discharge should be identified and prevented or equalized.

7. Is endogenous respiration occurring?

   Yes   See 8
   No   See 9

8. If possible, decrease MLVSS concentration. However, a decrease in MLVSS may limit the ability of the SBR to nitrify and adequately degrade cBOD, especially during cold wastewater temperature.

9. Is the SBR nitrifying and nitrification is not required?

   Yes   See 10
   No   See 11

10. Nitrification can be terminated or reduced by decreasing MLVSS concentration or time for the React Phase.

11. Low DO concentration may occur because of inadequate oxygen transfer. Check diurnal DO demand of the MLVSS to variations in organic loading and adjust or balance loading to match existing aeration.

Check List for the Identification of Operational Factors Responsible for Low Dissolved Oxygen

Operational Factor	√ If Monitored	√ If Possible Factor
Aeration equipment malfunction		
Oxygen scavengers		
Increase in organic loading		
Endogenous respiration		
Undesired nitrification		
Inadequate oxygen transfer		

Part IV

BNR and Phosphorus Removal

15

Nutrients

The nutrients most commonly found to be deficient in biological wastewater treatment processes are nitrogen and phosphorus. These nutrients perform many critical roles in cellular metabolism, especially enzymatic activity and energy transfer, cellular growth (sludge production), and protein synthesis. There are several techniques that can be used as guidelines to ensure that adequate nutrients are present in the mixed liquor to degrade BOD and provide for proper floc formation. These techniques involve sampling the influent and the mixed liquor at the end of the React Phase.

Grab samples may be collected for each waste stream during peak loading conditions when demand for nutrients by bacterial cells is greatest. Influent samples may be monitored for a 100:5:1 ratio for BOD:N:P or preferably soluble cBOD:NH_4^+-N:HPO_4^{2-}-P. Mixed-liquor samples may be monitored for the presence of a negative India ink reverse stain (Figure 15.1) or a positive India ink reverse stain (Figure 15.2) and undesired growth of filamentous organisms that have a common growth factor for a nutrient deficiency (Table 15.1). The mixed-liquor filtrate may be monitored for residual or target values of ≥ 1.0 mg/L NH_4^+-N or ≥ 3.0 mg/L NO_3^--N and ≥ 0.5 mg/L HPO_4^{2-}-P.

100:5:1 RATIO

The ratio of 100:5:1 indicates that for every 100 parts of total BOD there should be at least 5 parts of total nitrogen and at least 1 part of total phosphorus. The ratio is based upon the quantities of nitrogen and phosphorus that are needed to produce new bacterial cells (sludge) for every 100 parts of substrate (BOD) degraded by the

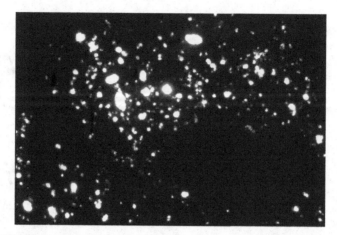

Figure 15.1   *Negative India ink reverse stain. Under an India ink reverse stain using a phase contrast microscope, bacterial cells stain black and/or golden-brown. Stored food or polysaccharides block the penetration of ink (carbon black particles) into the floc particle and therefore appears white. If the majority of the area of the floc particle is black and/or golden-brown rather than white, there is relatively little stored food in the floc particle and a low probability of a nutrient deficiency during React Phase in an SBR. This is a negative India ink reverse stain.*

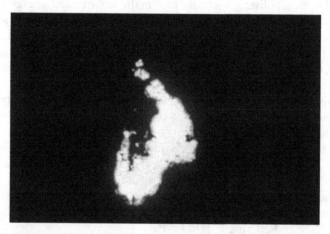

Figure 15.2   *Positive India ink reverse stain. Under an India ink reverse stain using a phase contrast microscope, bacterial cells stain black and/or golden-brown. Stored food or polysaccharides block the penetration of ink (carbon black particles) into the floc particle and therefore appears white. If the majority of the area of the floc particle is white rather than black and/or golden-brown, there is a relatively large quantity of stored food in the floc particle and a high probability of a nutrient deficiency during the React Phase in an SBR. This is a positive India ink reverse stain.*

bacteria. Because soluble cBOD is degraded first by bacteria, and bacteria use readily available nutrients such as NH_4^+-N for nitrogen and HPO_4^{2-} for phosphorus for the degradation of soluble cBOD, the ratio 100:5:1 is more applicable as soluble $cBOD:NH^{4+}$-N$:HPO_4^{2-}$-P.

TABLE 15.1 Filamentous Organisms and Common Growth Factor for a Nutrient Deficiency

Filamentous Organism	Growth Factors
Haliscomenobacter hydrossis	Low dissolved oxygen Low F/M (<0.05) **Low N or P**
Nocardioforms	Fats, oils and grease Low F/M (<0.05) **Low N or P** Low pH (<6.8) Slowly degradable substrates
Sphaerotilus natans	Low dissolved oxygen **Low N or P** Readily degradable substrates Warm wastewater temperature
Thiothrix spp.	**Low N or P** Organic acids Readily degradable substrates Septicity/sulfides
Type 0041	High MCRT (>10 days) Low F/M (<0.05) **Low N or P** Septicity/sulfides Slowly degradable substrates
Type 0092	High MCRT (>10 days) Fats, oils and grease Low F/M (<0.05) **Low N or P** Slowly degradable substrates
Type 0675	High MCRT (>10 days) Low F/M (<0.05) **Low N or P**
Type 1701	Low dissolved oxygen **Low N or P** Warm wastewater temperature
Type 021N	Low F/M (<0.05) **Low N or P** Organic acids Readily degradable substrates Septicity/sulfides

INDIA INK REVERSE STAIN

During a nutrient deficiency absorbed, soluble cBOD cannot be degraded properly and is converted to and stored as an insoluble polysaccharide outside the bacterial cells in the floc particles. The relative quantity of stored polysaccharide in floc particles can be observed by performing an India ink reverse stain (Table 15.2). During the staining technique the penetration of the carbon black particles in the ink into the floc particles is blocked by the polysaccharides. Where bacterial cells are present, they stain black or golden-brown under phase contrast microscopy. The polysaccharides do not stain and appear white under phase contrast microscopy. If most of the area of the floc particles observed is black/golden brown (Figure 15.1), there is

TABLE 15.2   Microscopic Technique for the India Ink Reverse Stain Solution: India Ink (aqueous solution of carbon black particles) or Nigrosine

Step	Direction
1	Mix one or two drops of India ink and one drop of mixed liquor on a microscope slide.
2	Place a cover slip on the India ink-mixed liquor sample and observed the sample at 1000× (oil immersion) under phase contrast microscopy.
3	Be sure that the floc particles that are being examined are surrounded by a black field of view.
4	In "nutrient-adequate" mixed liquor, the carbon black particles penetrate the floc particles almost completely, at most leaving only a few "spots" of white. This is a negative reaction to the India ink reverse stain.
5	In "nutrient-deficient" mixed liquor, large amounts of polysaccharides (produced through a nutrient deficiency) are present. The polysaccharides prevent the penetration of the carbon black particles. This results in the appearance of large white areas in the floc particles. This is a positive reaction to the India ink reverse stain.

a relatively small quantity of polysaccharides (stored food) and a low probability of a nutrient deficiency. However, if most of the area of the floc particles observed is white (Figure 15.2), there is a relatively large quantity of polysaccharides (stored food) and a high probability of a nutrient deficiency.

FILAMENTOUS ORGANISMS

There are approximately 30 filamentous organisms that are found in activated sludge processes. These organisms proliferate under a large variety of operational conditions. Several of these organisms have a common growth factor of a nutrient deficiency that permits their rapid and undesired growth over floc-forming bacteria. The filamentous organisms include *Haliscomenobacter hydrossis*, Nocardioforms, *Sphaerotilus natans*, *Thiothrix* spp., type 0041, type 0092, type 0675, type 1701, and type 021N (Table 15.1). The occurrence of these filamentous organisms in undesired numbers in the same mixed-liquor sample may be indicative of a nutrient deficiency.

TARGET VALUES

Ammonical-nitrogen is the preferred bacterial nutrient for nitrogen, and orthophosphate-phosphate is the preferred bacterial nutrient for phosphorus. Each of these nutrients is used first by bacterial cells for the degradation of soluble cBOD and the production of new bacterial cells (sludge). Therefore, if a mixed-liquor filtrate sample obtained at the end of React Phase contains a residual quantity or desired target value of ≥1.0 mg/L for ammonical-nitrogen and ≥0.5 mg/L for orthophosphate-phosphorus and the mixed liquor did not experience a toxic event, then the preferred nutrients were not exhausted and a nutrient deficiency did not occur during the React Phase.

If an SBR is required to completely nitrify, then the ammonical-nitrogen value would be <1 mg/L. In the absence of ammonical-nitrogen, bacteria use nitrate-nitrogen as their back-up nutrient for nitrogen. The presence of >3 mg/L nitrate-nitrogen and absence of toxicity also is indicative of an adequate quantity of readily available nitrogen.

If toxicity occurs in the mixed liquor, bacterial activity and growth is inhibited. Because growth is inhibited, there is no need for bacteria to degrade soluble cBOD or to absorb and use nutrients. Therefore, the quantity of nutrients in the mixed-liquor filtrate at the end of the React Phase as well as the quantity of soluble cBOD would be relatively high. The presence of toxicity may be detected by the occurrence of a significant decrease in the value for the specific oxygen uptake rate (SOUR).

INDUSTRIAL WASTEWATERS

Because soluble cBOD is degraded first in the mixed liquor, the largest demand for readily available nitrogen and phosphorus occurs when the largest quantity of soluble cBOD is present in the mixed liquor. Domestic wastewater (Figure 15.3)

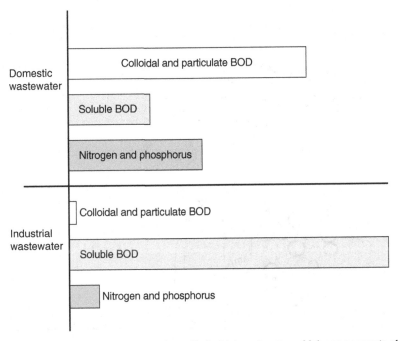

Figure 15.3 *BOD and nutrients in domestic and industrial wastewaters. Major components of domestic and industrial wastewaters with respect to the degradation of BOD consist of the types of BOD (colloidal, particulate, and soluble) and their relative abundance as well as the relative abundance of readily available nutrients (ammonical-nitrogen and orthophosphate-phosphorus). Domestic wastewater (top half of the bar graph) contains much colloidal and particulate BOD that degrades slowly; that is, it must undergo solubilization first before it can be absorbed and degraded. This mode of bacterial action does not place an immediate and significant demand on the SBR for dissolved oxygen and nutrients. In addition, domestic wastewater typically has a relatively large quantity of readily available nitrogen and phosphorus for bacterial use. With few exceptions, industrial wastewater has a relatively small quantity of colloidal and particulate BOD and a relatively large quantity of soluble BOD (bottom half of the bar graph). This relatively large quantity of easily and rapidly degradable cBOD places an immediate and large demand on the SBR for dissolved oxygen and readily available nutrients. Unfortunately, most industrial wastewaters are deficiency for readily available nitrogen (ammonical-nitrogen) and/or phosphorus (orthophosphate-phosphorus).*

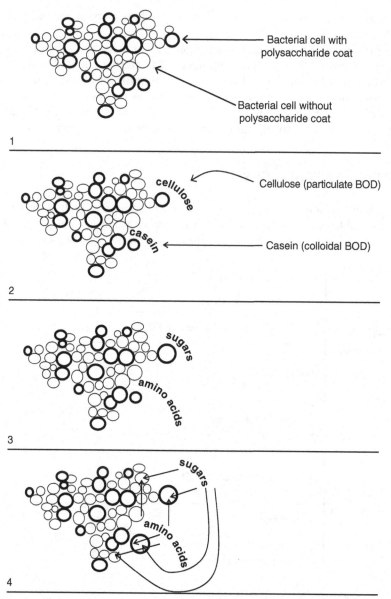

Labels in figure:

Bacterial cell with polysaccharide coat

Bacterial cell without polysaccharide coat

1

cellulose — Cellulose (particulate BOD)

casein — Casein (colloidal BOD)

2

sugars

amino acids

3

sugars

amino acids

4

Figure 15.4 *Adsorption and solubilization of colloidal BOD and particulate BOD. There are two basic types of bacterial cells of concern with respect to the removal and degradation of colloidal BOD and particulate BOD. These cells are those that possess a polysaccharide coat and those that do not possess a polysaccharide coat (1). Colloidal BOD such as casein and particulate BOD such as cellulose are adsorbed to the surface of bacterial cells that have a polysaccharide coat (2). Once the colloidal BOD and particulate BOD have been adsorbed, the polysaccharide-coated cells produce exoenzymes that leave the cell and "attack" the adsorbed substrates (casein and cellulose). The exoenzymes convert the complex and insoluble substrates into simple and soluble substrates. Casein is converted to simple amino acids, and cellulose is converted to simple sugars (3). These simple substrates (amino acids and sugars) can then be absorbed by bacterial cells with and without a polysaccharide coating where they degraded inside the cells by endoenzymes (4).*

TABLE 15.3   Nutrient Deficient Industrial Wastewaters

Industrial Wastewater	Nitrogen Deficient	Phosphorus Deficient
Bakery	X	
Beverage, malt	X	X
Beverage, distilled spirits	X	X
Beverage, wine	X	X
Beverage, soda drink	X	X
Citrus		X
Coffee	X	X
Coke ovens		X
Corn	X	
Cotton kerning	X	
Dairy, milk		X
Dairy, cottage cheese	X	
Food processing	X	X
Formaldehyde	X	X
Fruit and vegetable	X	X
Leather tanning		X
Petroleum		X
Pharmaceutical		X
Phenols	X	
Pulp and paper	X	X
Textile	X	

contains a relatively large quantity of colloidal BOD and particulate BOD that are adsorbed to floc particles and slowly solubilized (Figure 15.4). Therefore, the presence of domestic wastewater in the SBR does not exert an immediate and significant demand for nutrients. Also, domestic wastewater has a relatively large quantity of readily available nitrogen and phosphorus. However, most industrial wastewaters (Figure 15.3) contain a relatively large quantity of soluble cBOD that exerts an immediate and significant demand for readily available nutrients when they are present in the SBR. Unfortunately, many industrial wastewaters (Table 15.3) are deficient for readily available nutrients. If nutrient needs are not satisfied in an SBR, then operational problems including the following may occur:

- Loss of treatment efficiency
- Production of billowy white foam or greasy gray foam
- Production of slimy floc that interferes with desired settleability of solids and dewatering of solids
- Undesired growth of nutrient deficient filamentous organisms

To correct for a nutrient deficiency, the deficient nutrient or nutrients must be identified and added either to the industrial wastewater or the SBR. Table 15.4 provides a list of chemical compounds that may be used to correct for a nutrient deficiency.

Often nutrients are found in higher quantities than needed for bacterial growth—that is, assimilation of nitrogen and phosphorus in new bacterial cells or sludge. The excess nutrients unfortunately are discharged from the SBR to receiving

TABLE 15.4   Chemical Compounds Suitable for Nutrient Addition

Nutrient Needed	Chemical Compound	
	Common Name	Formula
Nitrogen	Anhydrous ammonia	NH_3
	Aqua ammonia	NH_4OH
	Ammonium bicarbonate	NH_4HCO_3
	Ammonium carbonate	$(NH_4)_2CO_3$
	Ammonium chloride	NH_4Cl
	Ammonia sulfate	$(NH_4)_2SO_4$
Phosphorus	Trisodium phosphate	Na_3PO_4
	Disodium phosphate	Na_2HPO_4
	Monosodium phosphate	NaH_2PO_4
	Sodium hexametaphosphate	$Na_3(PO)_6$
	Sodium tripolyphosphate	$Na_5P_3O_{10}$
	Tetrasodium pyrophosphate	$Na_4P_2O_7$
	Phosphoric acid	H_3PO_4
Nitrogen and/or phosphorus	Ammonium phosphate	$NH_4H_2PO_4$

waters. These discharges contribute to pollution problems that affect the quality of the receiving waters and the quality of potable water obtained from the receiving water. Therefore, nutrient removal mechanisms besides nutrient assimilation must be used at wastewater treatment plants to reduce the quantity of nutrients discharged.

16

Biological Nutrient Removal

Nitrogen and phosphorus are the major nutrients that are responsible for cultural eutrophication of surface waters. Although algal blooms are the most commonly occurring and most easily recognizable consequence of eutrophication, there are many significant and undesired consequences of eutrophication. These undesired consequences or adverse impacts upon water quality and human health include:

- Low dissolved oxygen level in the surface water as a result of the death and decomposition of algae and other aquatic plants
- Low dissolved oxygen level in the surface water as a result of the oxidation (nitrification) of ammonium (NH_4^+) to nitrate (NO_3^-)
- Undesired turbidity due to the growth, death and decomposition of flora and fauna
- Undesired chlorine demand to disinfect surface water for a potable water supply
- Production of carcinogenic or potentially carcinogenic compounds from the increased chlorination of the surface water when used as a potable water supply and
- Occurrence of methemoglobinemia ("blue baby" disease) from the ingestion of nitrate-laden surface water when used as a potable water supply.

NITROGEN REMOVAL

Total effluent nitrogen consists of ammonium (NH_4^+), nitrite (NO_2^-), nitrate (NO_3^-), and organic-nitrogen (Figure 16.1). Organic-nitrogen is present in a waste

Troubleshooting the Sequencing Batch Reactor, by Michael H. Gerardi
Copyright © 2010 by John Wiley & Sons, Inc.

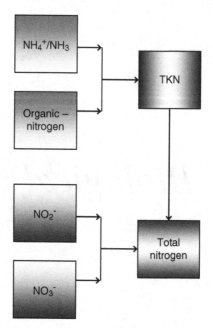

Figure 16.1 *Total effluent nitrogen. Total nitrogen in the effluent consists of total Kjeldahl nitrogen (TKN), nitrite (NO₂⁻), and nitrate (NO₃⁻). TKN consists of ammonium (NH₄⁺)/ammonia (NH₃) and organic nitrogen.*

stream as amino acids, proteins, polymers, and surfactants and exists in the soluble and insoluble forms. In the soluble form, bacterial cells absorb it; in the insoluble form, bacterial cells and inert solids adsorb it (Figure 16.2).

Nitrogen is removed from a waste stream through several biological, chemical, and physical processes. A single process or combination of process can remove nitrogen. These processes may be grouped as follows:

- Stripping of nitrogen to the atmosphere as gaseous ammonia (NH_3).
- Assimilation and adsorption of nitrogen by bacterial cells and inert solids.
- Ammonification, nitrification, and denitrification of nitrogen to gaseous molecular nitrogen (N_2).

STRIPPING OF NITROGEN

Reduced nitrogen exists in the mixed liquor as ammonium (NH_4^+) and ammonia (NH_3). Both forms of reduced nitrogen contain hydrogen, and the dominant form is pH-dependent (Figure 16.3). With increasing pH, ammonium is converted to ammonia and some ammonia is "stripped" or lost to the atmosphere through aeration and mixing action. Air stripping of ammonia removes nitrogen from a waste stream. However, only a relatively small quantity (<10%) of ammonium is converted to gaseous ammonia and stripped to the atmosphere with pH fluctuations in the mixed liquor.

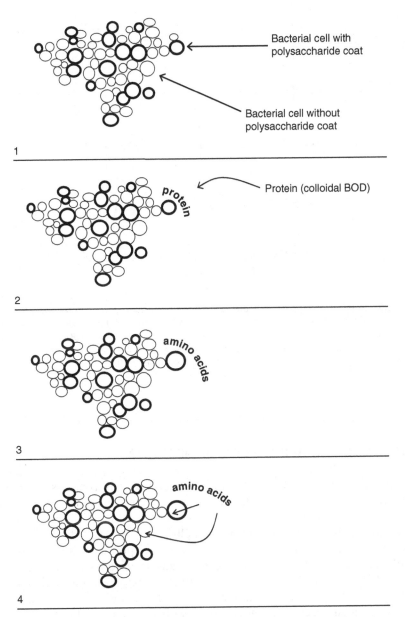

Figure 16.2 *Adsorption and solubilization of nitrogenous compounds. Nitrogenous compounds in the colloidal state such as proteins are adsorbed by the surface of bacterial cells (1) that possess a polysaccharide coat (2). Once adsorbed, the cell produces and releases an exoenzyme that "attacks" and solubilizes the protein to produce simple and soluble amino acids (3). The soluble amino acids are then absorbed by bacterial cells that do and do not have a polysaccharide coat where they are degraded inside the cell by endoenzymes (4).*

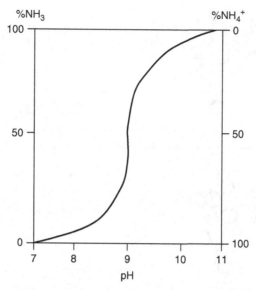

Figure 16.3 *Relative abundance of reduced nitrogen (NH₃ and NH₄⁺). The relative abundance or amount of reduced nitrogen as ammonia (NH₃) and ammonium (NH₄⁺) is dependent upon the pH of the SBR. With increasing pH ammonia increases in quantity and conversely with decreasing pH ammonium increases in quantity. Shown is the change in the dominant form of reduced nitrogen at pH values of <9.4 (ammonium favored) and ≥9.4 (ammonia favored).*

ASSIMILATION AND ADSORPTION OF NITROGEN

Nitrogen is assimilated or incorporated into bacterial cells as an essential nutrient for the synthesis of cellular material for the production of new cells (sludge). Nitrogen as ammonium and nitrate are absorbed from the bulk solution and used by bacterial cells for the synthesis of cellular material and cellular growth. Approximately 15% of the dry weight of new sludge produced is nitrogen.

Insoluble organic-nitrogen compounds are adsorbed to bacterial cells and inert solids in floc particles. If adequate retention time occurs, organic-nitrogen compounds may undergo ammonification that results in the release of ammonium. Soluble organic-nitrogen compounds are absorbed by bacterial cells and undergo ammonification.

Although nitrogenous compounds are assimilated by bacterial cells and adsorbed by bacterial cells and inert solids, they are not removed from the waste stream until the nitrogen-laden bacterial cells and inert solids have been wasted. Wasted bacterial cells and inert solids that undergo additional treatment may release nitrogen back to the mixed liquor through recycle streams, especially through aerobic digesters (Figure 16.4) and anaerobic digesters (Figure 16.5) and sludge dewatering processes. Often, these recycle streams contain polymers such as polyacrylamides polymers that are organic-nitrogen in composition and release ammonium upon their degradation.

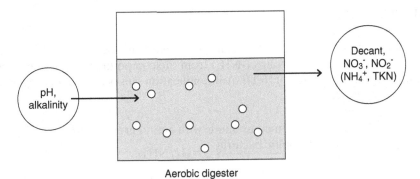

Aerobic digester

Figure 16.4 *Release of nitrogenous compounds from an aerobic digester. Solids transferred to a properly operating aerobic digester undergo ammonification (release of amino groups) and nitrification. Nitrification results in the production of nitrate (NO_3^-) and a decrease in digester alkalinity and pH. Therefore, the digester should be monitored periodically to ensure proper pH and alkalinity. The decant from the digester should be monitored to determine recycle of nitrogenous compounds. These compounds would include nitrate, nitrite (NO_2^-), ammonium (NH_4^+), and TKN. These compounds also should be monitored in the centrate/filtrate of dewatering operations.*

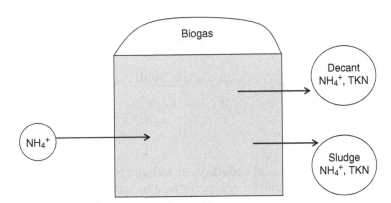

Figure 16.5 *Release of nitrogenous compounds from an anaerobic digester. Solids transferred to a properly operating anaerobic digester undergo ammonification. Because free molecular oxygen is not available in an anaerobic digester, nitrification does not occur. Therefore, reduced nitrogen (NH_3/NH_4^+) accumulates in the digester. At elevated concentrations of ammonia ($\geq 1500\,mg/L$), ammonia toxicity may occur. The decant from the digester should be monitored to determine recycling of nitrogenous compounds (ammonium and TKN). These compounds also should be monitored in the centrate/filtrate of dewatering operations.*

AMMONIFICATION, NITRIFICATION, AND DENITRIFICATION

Collectively, ammonification, nitrification, and denitrification are responsible for the removal of significant quantities of nitrogen from a waste stream. These biological processes permit compliance with a total nitrogen discharge requirement. Of these three processes, nitrification is the most difficult to achieve. Denitrification is the process that removes nitrogen from the waste stream as molecular nitrogen (N_2) and returns it to the atmosphere. Ammonification and nitrification only change the form of nitrogen in the waste stream.

AMMONIFICATION

Ammonification is the bacterial degradation of organic-nitrogen compounds, resulting in the release of amino groups (-NH$_2$) from the compounds and the production of ammonia/ammonium [Eq. (16.1)]. Ammonification can occur under aerobic and anaerobic conditions.

$$\text{Organic-nitrogen compounds with amino groups}\,(\text{-NH}_2) \xrightarrow{\text{organotrophic bacteria}} \text{NH}_3/\text{NH}_4^+ \tag{16.1}$$

NITRIFICATION

Nitrification is a two-step biochemical reaction involving the oxidation (addition of oxygen) to ammonium (NH$_4^+$) to nitrite (NO$_2^-$) [Eq. (16.2)] and then the oxidation of nitrite to nitrate (NO$_3^-$) [Eq. (16.3)]. Ammonia-oxidizing bacteria (AOB) such as *Nitrosomonas* are responsible for the oxidation of ammonium, and nitrite-oxidizing bacteria (NOB) such as *Nitrobacter* are responsible for the oxidation of nitrite.

$$\text{Step 1:}\quad \text{NH}_4^+ + 1.5\text{O}_2 \xrightarrow{\text{AOB}} \text{NO}_2^- + 2\text{H}^+ + \text{H}_2\text{O} \tag{16.2}$$

$$\text{Step 2:}\quad \text{NO}_2^- + 0.5\text{O}_2 \xrightarrow{\text{NOB}} \text{NO}_3^- \tag{16.3}$$

DENITRIFICATION

Denitrification is the biological reduction of nitrate (NO$_3^-$) to gaseous molecular nitrogen [Eq. (16.4)]. Denitrification is performed by facultative anaerobic, organotrophic bacteria or denitrifying bacteria such as *Bacillus* and *Pseudomonas*. Denitrification occurs in the presence of soluble cBOD and the absence of free molecular oxygen (O$_2$). Denitrification removes nitrogen from the waste stream as insoluble molecular nitrogen.

$$\text{NO}_3^- + \text{soluble cBOD} \xrightarrow{\text{denitrifying bacteria}} \text{N}_2 + \text{CO}_2 + \text{OH}^- + \text{H}_2\text{O} \tag{16.4}$$

PHOSPHORUS REMOVAL

Total effluent phosphorus consists of soluble and insoluble forms. Phosphorus is removed from the waste stream through biological and chemical/physical processes. The biological processes consist of (1) assimilation and adsorption of phosphorus and (2) enhanced biological phosphorus removal (EBPR) or "luxury uptake" of phosphorus. Phosphorus can also be removed from a waste stream through chemical reaction with a precipitant such as a coagulant (metal salt) or polymer and then wasted from the treatment system.

ASSIMILATION AND ADSORPTION OF PHOSPHORUS

Phosphorus is assimilated or incorporated into bacterial cells as an essential nutrient for (1) energy storage, (2) cellular activity, (3) synthesis of cellular material, and (4) cellular growth (sludge production). Soluble orthophosphate or reactive phosphorus is used by bacterial cells as the phosphorus nutrient for cellular needs. In the mixed liquor, orthophosphate exists in two forms, $H_2PO_4^-$ and HPO_4^{2-} (Figure 16.6). The dominant form of orthophosphate is pH-dependent. Approximately 3% of the dry weight of bacterial cells or sludge is phosphorus.

Insoluble phosphorous compounds are adsorbed to bacterial cells and inert solids in floc particles. If adequate retention time occurs, these compounds may be degraded or hydrolyzed resulting in the release of orthophosphate. Bacteria hydrolyze complex, insoluble phosphorous compounds. Hydrolysis of these compounds results in the release of orthophosphate.

Phosphorous compounds that are assimilated by bacterial cells and adsorbed by bacterial cells and inert solids are removed from the waste stream when the bacterial cells and inert solids are wasted. However, wasted bacterial cells and inert solids that contain phosphorous compounds and undergo additional treatment may release phosphorus back to the treatment system through recycle streams.

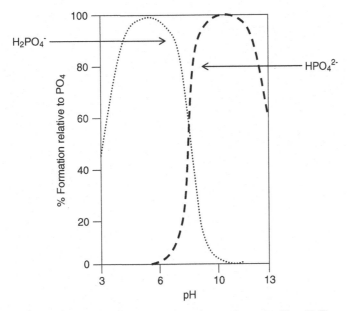

Figure 16.6 *Forms of orthophosphate in the mixed liquor as determined by pH. There are two forms of reactive phosphorus or orthophosphate that are found in the mixed liquor at typical operating pH values (6.8–7.2). These forms are $H_2PO_4^-$ and HPO_4^{2-}. $H_2PO_4^-$ is dominant at pH values of <7, while HPO_4^{2-} is dominant at pH values of >7.*

ENHANCED BIOLOGICAL PHOSPHORUS REMOVAL

Enhanced biological phosphorus removal or EBPR is dependent upon the uptake and assimilation of orthophosphate by aerobic, organotrophic poly-P bacteria or phosphorus accumulating organisms (PAO) such as *Acineobacter* in excess of their biological growth requirements. The excess or "luxury uptake" of phosphorus is the result of cycling the poly-P bacteria between anaerobic/fermentative (Static Fill Phase) and aerobic (React Phase) conditions. The anaerobic/fermentative condition results in biological phosphorus release, while the aerobic condition results in biological phosphorus ("luxury") uptake. When phosphorus-rich bacteria (sludge) are wasted from the treatment process, a relatively large quantity of phosphorus also is wasted from the treatment process. Typically, bacterial cells (sludge) contain approximately 3% phosphorus by dry weight in treatment processes that do not practice enhanced biological phosphorus removal, while bacterial cells (sludge) contain approximately 7% phosphorus by dry weight in treatment processes that do practice enhanced biological phosphorus removal.

17

Chemical Phosphorus Removal

Excess phosphorus in surface waters often is responsible for the rapid and undesired growth of aquatic plants including algae. The growth and death of the plants contributes to numerous pollution problems including eutrophication. These problems are being addressed by regulatory agencies in part by placing stringent nitrogen and phosphorus discharge requirements on wastewater treatment plants. Although phosphorus discharge can be reduced through biological phosphorus removal (enhanced biological phosphorus removal or luxury uptake of phosphorus), biological phosphorus removal can be difficult to control. Also, reducing the effluent total phosphorus to <2.0 mg/L is difficult to achieve through biological phosphorus removal. Chemical precipitation of phosphorus following wastewater treatment can reduce effluent total phosphorus to <2.0 mg/L.

Total phosphorus in domestic wastewater typically is 6–8 mg/L. This consists of approximately 3–4 mg/L of orthophosphate or reactive phosphorus, 2–3 mg/L of condensed phosphorus such as polyphosphates (Table 17.1), and 1 mg/L of organic-phosphorus (Table 17.2). Condensed phosphorus and organic phosphorus can be degraded or converted to reactive phosphorus through bacterial degradation and hydrolysis, respectively, in an SBR.

Secondary wastewater treatment systems such as the SBR typically remove 1–2 mg/L of phosphorus. This relatively low quantity of phosphorus removal is due to its incorporation or assimilation in new bacterial cells or sludge. Currently, many discharge requirements for phosphorus are <2 mg/L. In order to satisfy these requirements, chemical precipitation is often used to remove orthophosphate, an inorganic form of phosphorus. The multivalent ions most commonly used for phosphorus precipitation are calcium (Ca^{2+}), aluminum (Al^{3+}), and iron (Fe^{2+} and Fe^{3+}).

Phosphorus removal in a conventionally operated SBR occurs through its adsorption to floc particles and its assimilation into new cells (sludge) as the

Troubleshooting the Sequencing Batch Reactor, by Michael H. Gerardi
Copyright © 2010 by John Wiley & Sons, Inc.

TABLE 17.1 Examples of Condensed Phosphates or Polyphosphates

Chemical Name	Formula	Structure
Diphosphate (pyrophosphate)	$H_4P_2O_7$	Linear
Triphosphate	$H_5P_3O_{10}$	Linear
Tetraphosphate	$H_6P_4O_{13}$	Linear
Trimetaphosphate	$(HPO_3)_3$	Cyclic
Tetrametaphosphate	$(HPO_3)_4$	Cyclic

TABLE 17.2 Sources of Significant Phosphorus

Fecal waste
Industrial and commercial waste
Synthetic detergents and cleaning products

TABLE 17.3 Chemical Precipitants for Phosphorus Removal

Chemical Precipitant	Common Name	Formula
Aluminum sulfate	Alum	$(Al_2(SO_4)_3 \cdot 16H_2O$
Calcium hydroxide	Hydrated lime	$Ca(OH)_2$
Ferric chloride	Iron chloride	$FeCl_3$
Ferrous sulfate	Iron sulfate	$FeSO_4$

phosphorus nutrient. Conventionally operated SBR do not practice biological phosphorus removal (enhanced biological phosphorus removal or luxury uptake of phosphorus).

Orthophosphate is readily available for biological absorption and assimilation without further breakdown. Orthophosphate also is the reactive form of phosphorus (reactive phosphorus) and the only form of phosphorus that can be precipitated by aluminum, calcium and iron based precipitants. Organic phosphate contains carbon and hydrogen and cannot be removed from the waste stream by chemical precipitation. However, it is adsorbed by floc particles and hydrolyzed over time by bacterial activity. Once hydrolyzed, organic phosphate releases orthophosphate. Polyphosphates are chemical compounds with two or more phosphorus atoms. Although polyphosphates usually undergo hydrolysis and revert to the orthophosphate form, polyphosphate hydrolysis is usually slow.

Adsorption and assimilation of phosphorus by floc particles and bacteria in an SBR removes approximately 30% of influent phosphorus from domestic wastewater. This leaves approximately 4–5 mg/L of total phosphorus in the decant. Chemical precipitation of phosphorus can reduce the total phosphorus in the decant to <1 mg/L. However, it is necessary to waste or convert influent condensed phosphorus and organic phosphorus to reactive phosphorus in order to reduce total phosphorus in the decant to this value. For example, if the decant total suspended solids (TSS) is ≥20 mg/L with a phosphorus content of 5% (50 mg P/g SS), a total phosphorus concentration of <1 mg/L cannot be achieved.

Removal of precipitated phosphorus is dependent upon the addition of a chemical precipitant (Table 17.3) and adjustment in pH. The chemical precipitant (coagulant or metal salt) of choice often is aluminum sulfate ($Al_2(SO_4)_3 \cdot 16H_2O$. Aluminum sulfate is commonly used for phosphorus removal in wastewater with high alkalinity and low stable phosphorus concentrations.

Aluminum sulfate addition requires flash mixing or high turbulence, flocculation, sedimentation, and solids removal. The addition of aluminum sulfate results in the production of insoluble aluminum phosphate ($AlPO_4$) [Eq. (17.1)], decrease in alkalinity and pH, and increase in sludge production.

$$Al_2(SO_4)_3 \cdot 16H_2O + 2PO_4^{-3} \rightarrow 2AlPO_4 + 3SO_4^{2-} + 16H_2O \qquad (17.1)$$

The optimum pH for phosphorus precipitation with aluminum sulfate is 6.2, and the optimum range for phosphorus precipitation is 5.5–6.5. At optimum pH a react time of approximately 2 hr is required for aluminum phosphate production with the addition of alum. This react time is based upon iron chemistry for the precipitation of phosphorus. Precipitation of phosphorus using aluminum or iron salts has similar chemistry. A longer react time is required at pH values greater than or less than the optimum pH. However, recent advances in chemical injection and mixing technologies considerably shorten react time.

Aluminum phosphate (density 2.566) has a good settling rate. The addition of 40 mg/L of aluminum sulfate produces a floc that settles at a rate of approximately 380 mm/min (0.124 ft/min). However, the floc may be easily resuspended. Therefore, an anionic polymer may be added to increase floc size and density and permit the floc to settle move quickly.

Approximately 5.8 mg of alkalinity as calcium carbonate ($CaCO_3$) are consumed per milligram of aluminum removed, and 2.9 mg of solids are produced per milligram of aluminum removed. Chemical precipitation of phosphorus may increase sludge production by 10% to 25%. Sludge production from the addition of aluminum sulfate is difficult to dewater due to the presence of entrapped water.

Aluminum sulfate may be added in liquid or dry form and may be added to (1) raw or influent wastewater, (2) the mixed liquor (simultaneous precipitation or phosphorus), (3) the decant or phosphorus precipitation tank downstream of the SBR (post precipitation of phosphorus), or (4) mixed liquor and decant (multiple precipitation of phosphorus). Approximately 9.6 mg of aluminum sulfate are required for each milligram of phosphorus removed to achieve total phosphorus decant concentration of approximately 0.5 mg/L. To achieve total phosphorus decant concentrations of <0.5 mg/L, the quantity of aluminum sulfate needed may be twice the value typically used due to competing chemical reactions. The main competition for aluminum sulfate is the quantity of bicarbonate alkalinity (HCO_3^-) in the wastewater [Eq. (17.2)]. This competition results in the production of aluminum hydroxide ($Al(OH)_3$).

$$Al_2(SO_4)_3 \cdot 16H_2O + 6HCO_3^- \rightarrow 2Al(OH)_3 + 6CO_2 + 16H_2O + 3SO_4^{-3} \qquad (17.2)$$

Post precipitation of phosphorus is a standard mode of operation. Post precipitation provides the highest phosphorus removal efficiency and can produce a final effluent phosphorus value of <0.5 mg/L. Post precipitation downstream of the SBR also removes many suspended solids that escape the SBR in the decant.

The quantity of aluminum sulfate necessary for phosphorus removal is better determined through full-scale evaluation rather than jar testing. It is typical to overdose aluminum sulfate at average influent phosphorus concentration.

Simultaneous precipitation of phosphorus has the following advantage and disadvantage:

Advantage. Precipitated phosphorus is incorporated into the sludge and improves settleability.

Disadvantage. Decrease in alkalinity and pH due to aluminum sulfate addition may inhibit floc formation, nitrification and enzymatic activity.

If an SBR is used for nitrification and denitrification, simultaneous precipitation of phosphorus may be difficult to achieve.

Post precipitation of phosphorus in an SBR has the following advantage and disadvantages:

Advantage. No adverse impact upon mixed liquor biological activity.

Disadvantage. Lowers alkalinity and pH and may cause pH violations of discharge permit limits, unless pH is adjusted before discharge. Sodium hydroxide (NaOH) often is used to adjust low, final effluent pH.

Disadvantage. Lowers alkalinity and pH and may interfere with disinfection of the effluent via chlorination with aqueous chlorine (Cl_2).

The addition of sodium hypochlorite (NaOCl) and calcium hypochlorite ($Ca(OCl)_2$) to wastewater for disinfection yields the hypochlorite ion (OCl^-) at an elevated pH. Hypochlorite raises the pH of the wastewater. An alkali is typically added to sodium hypochlorite and calcium hypochlorite to enhance the stability of the compounds and slow their decomposition. Thus, post precipitation of phosphorus may not adversely affect hypochlorite disinfection of the effluent.

Enhance flocculation of precipitated phosphorus and solids may be achieved with anionic polymer addition. Polymer addition helps to prevent pin floc and carryover of solids. Approximately 0.1–1.0 mg/L of polymer typically is required to enhance flocculation. The appropriate anionic polymer for use should be determined through jar testing, and the polymer should be added with good mixing action and low turbulence. The anionic polymer may decrease pH.

Poly-aluminum chloride (PAC) refers to a group of soluble aluminum compounds in which aluminum chloride ($AlCl_3$) has been reacted with a base or hydroxyl ion (OH^-). The relative amount of hydroxide ions compared to the amount of aluminum determines the basicity of a particular PAC product. Because PAC is less acidic than alum, there is less risk of shocking the treatment system with a pH change when PAC is used to chemically precipitate phosphorus.

18

Biological Phosphorus Removal

The goal of biological phosphorus removal is to incorporate as much phosphorus as possible into bacterial cells (MLVSS or sludge). In conventional activated sludge processes approximately 1% to 3% of the dry weight of the waste activated sludge (WAS) contains phosphorus, while 3% to 7% of the dry weight of the WAS of bacterial cells or sludge contains phosphorus in activated sludge processes that utilize biological phosphorus removal. Biological phosphorus removal is also known as "enhanced biological phosphorus removal" (EBPR) and "luxury uptake of phosphorus." Biological phosphorus removal usually is achieved in SBR with a MLSS of 2000–3000 mg/L and a React Phase of approximately 30–60 min.

Biological phosphorus removal requires the use of two groups of bacteria and two operational conditions. Bacteria include the acid-forming or fermentative bacteria and the poly-P bacteria or phosphorus-accumulating organisms (PAO). These bacterial groups enter the SBR through fecal waste and inflow and infiltration (I/I) as soil and water organisms. The operational conditions include an anaerobic/fermentative period (Static Fill Phase) followed by an oxic or aerobic period (React Phase).

During the anaerobic/fermentative period or Static Fill Phase the following sequence of events occurs (Figure 18.1):

- Fermentative bacteria degrade soluble cBOD and produce fatty acids (Table 18.1) that are used as a carbon and energy source by the poly-P bacteria. Acetate is the major fatty acid that promotes biological phosphorus release by poly-P bacteria.
- Poly-P bacteria absorb but do not degrade the fatty acids during the anaerobic/fermentative period. The absorbed fatty acids are converted to and stored as

---

Troubleshooting the Sequencing Batch Reactor, by Michael H. Gerardi
Copyright © 2010 by John Wiley & Sons, Inc.

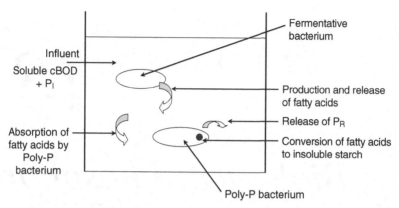

Figure 18.1 Biological events during the Static Fill Phase (anaerobic/fermentative period). During the static fill phase, free molecular oxygen and nitrates are not present or their residual quantities are depleted quickly by the bacteria in the SBR. In the absence of aeration and mixing, fresh influent enters the SRB. Under an anaerobic/fermentative condition, biological phosphorus release occurs. Fermentative (acid-forming) bacteria degrade the influent soluble cBOD and produce fatty acids. The acids are quickly absorbed by Poly-P bacteria. The Poly-P bacteria cannot degrade the fatty acids in the absence of free molecular oxygen and convert the fatty acids to insoluble starch granules. This conversion process requires the use of energy by the bacteria cells or the breakage of high-energy phosphate bonds. When this happens, the bacterial cells release orthophosphate (P_R) to the bulk solution. At the completion of the Static Fill Phase, the bulk solution contains two pools of orthophosphate, namely, the influent orthophosphate (P_I) and the released orthophosphate (P_R).

TABLE 18.1   Examples of Volatile Fatty Acids

Volatile Fatty Acid	Formula
Acetate	CH_3COOH
Butyrate	$CH_3(CH_2)_2COOH$
Formate	$HCOOH$
Lactate	$CH_3CHOHCOOH$
Propionate	CH_3CH_2COOH

insoluble starch granules or β-polyhydroxybuytrate (PHB) granules inside the bacterial cell.

• Poly-P bacteria expend energy to convert the fatty acids to insoluble starch. Energy expenditure results in the breakage of high-energy phosphate bonds inside the cell and the release of orthophosphate to the bulk solution. The release of orthophosphate is known as "biological phosphorus release."

• As a result of biological phosphorus release by the poly-P bacteria, the concentration of orthophosphate in the bulk solution increases. The bulk solution contains orthophosphate associated with the influent and orthophosphate released by the poly-P bacteria.

In addition to an increase in orthophosphate concentration during the Static Fill Phase, the concentration of alkalinity decreases and the pH drops due to the production of fatty acids. Also, a drop in ORP occurs during the Static Fill Phase.

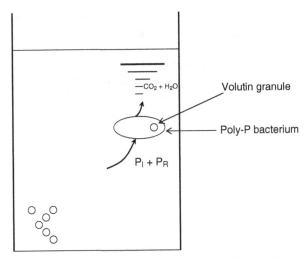

Figure 18.2 *Biological events during React Phase. During the React Phase that has been preceded by the Static Fill Phase, Poly-P bacteria solubilize the starch granules produced during the Static Fill Phase. The degradation of the starch granules results in the release of much energy from the freed electrons in the chemical bonds of the cBOD. The released energy is stored by the Poly-P bacteria in high-energy bonds to form volutin granules. Due to the large demand for phosphorus to store the released energy, large quantities of phosphorus from the influent (P_I) and the release phosphorus (P_R) during the Static Fill Phase are removed from the bulk solution.*

During the aerobic period or React Phase of an SBR, the following sequence of events occurs (Figure 18.2):

- Starch granules in the poly-P bacteria that were produced during the Static Fill Phase are solubilized and degraded.
- The degradation of solubilized starch granules results in the release of consider-able usable energy for bacterial cells.
- Poly-P bacteria remove large quantities of orthophosphate from the bulk solu-tion—associated with the influent and release during the Static Fill Phase—in order to store the released energy as high-energy phosphate bonds. The stored energy is package in insoluble polyphosphate granules or volutin granules. Phosphorus has been removed from the bulk solution in higher quantities than needed by bacteria for normal cellular activity, that is, luxury uptake of phosphorus has occurred.

The removal of orthophosphate by poly-P bacteria is known as "biological phos-phorus removal" and results in (1) a significant decrease in orthophosphate concen-tration in the bulk solution and decant and (2) a significant increase in phosphorus content in poly-P bacteria or sludge (MLVSS).

The major advantages of biological phosphorus removal as compared to chemical phosphorus precipitation are (1) reduced chemical costs and (2) decreased sludge production. Chemicals are not required for biological phosphorus removal, and a phosphorus-precipitated sludge is not produced. Depending on the chemical used to precipitate orthophosphate from the bulk solution, the phosphorus-precipitated sludge may be aluminum phosphate ($AlPO_4$), calcium phosphate ($Ca_3(PO_4)_2$), ferric phosphate ($FePO_4$), or ferrous phosphate ($Fe_3(PO_4)_2$). Also, chemical precipitation of orthophosphate may destroy alkalinity and lower pH.

Part V

Monitoring

19

Phases and Parameters

Samples of wastewater or mixed liquor may be collected and analyzed in the laboratory or *in situ* to monitor treatment efficiency, ensure cost-effective operation and permit compliance, and provide for proper process control and troubleshooting decisions. In order to make process control and troubleshooting decisions, representative sampling points and samples are required.

Samples are selected and analyzed based upon permit requirements and control parameters that are best for successful operation of the SBR. Several tests are usually required for permit compliance, while many tests usually are recommended to provide for process control and troubleshooting decisions (Table 19.1). In addition to the data obtained from the tests listed in Table 19.1, a range of values for process control parameters such as F/M, MCRT, MLSS, MLVSS, settleability, and sludge volume index (SVI) are critically important. A range in values rather than a specific value for each control parameter provides for a safe, steady-state operational condition and an acceptable quality of decant.

The range in values selected for a parameter may differ from literature values, warm and cold wastewater operational conditions, and industrial discharges. Each treatment process must develop its own parameters and range in values for each parameter that are most reflective of a successful steady-state operational condition.

For sample collection and sample testing, an operator must determine the appropriate location, type of sample, and time of sample. Locations would include center, mid-point, and perimeter of the SBR as well as depth of sample collection. Types of samples would include grab and composite samples and filtered and nonfiltered samples. Time of sample collection would include the beginning, mid-point, and end of a specific phase to determine (1) if the microbial event of the phase does occur

TABLE 19.1   Recommended Process Control and Troubleshooting Tests

Test	Comment[a]
Alkalinity (mg/L) (as CaCO$_3$)	Filtered sample
Ammonical-nitrogen	Filtered sample
BOD, mg/L and/or COD (mg/L)	Nonfiltered sample
Conductivity (μmohms/cm)	*In situ*
Dissolved oxygen (mg/L)	*In situ*
Nitrate-nitrogen (mg/L)	Filtered sample
Nitrite-nitrogen (mg/L)	Filtered sample
Oxidation–reduction potential (mV)	*In situ*
pH (standard units)	*In situ*
Orthophosphate-phosphorus (mg/L)	Filtered sample
Phosphorus, total (mg/L)	Nonfiltered sample
Sulfide (mg/L)	Filtered sample
Temperature (°C)	*In situ*
TKN (mg/L)	Nonfiltered sample

[a]To obtain a filtered sample (filtrate), the sample of wastewater or mixed liquor should be filtered through a Whatman® No. 4 filter paper.

(for example, is the Mix Fill Phase really anoxic?) and (2) the progress of the phase (for example, how rapidly are nitrates removed during the Mix Fill Phase?).

For process control and troubleshooting, tests are more meaningful and reflective of changes in the SBR if the samples are collected at the same location and time each day. By doing so, readings from different days can be compared. Obtained values are helpful in determining if existing operational conditions are satisfactory, including the duration of time for each phase.

In order to determine if a phase of operation of an SBR is performing properly and is meeting its specific goal, then it would be helpful to sample or monitor *in situ* the SBR at the beginning, mid-point, and end of the phase. Collected samples would be analyzed for specific parameters that are pertinent for the phase, while *in situ* monitoring would also test the mixed liquor for specific parameters. The amount of sampling and testing to be performed on an SBR depends upon the mode of operation of the SBR, the discharge requirements for the treatment process, the availability of manpower, and laboratory equipment and cost. Phases that should be monitored for acceptable performance include the following:

- Static Fill Phase
- Mix Fill Phase
- Aerated Fill Phase
- React Phase
- Carbon-fed denitrification period after the Settle Phase

STATIC FILL PHASE

The Static Fill Phase is an anaerobic/fermentative condition. If a Static Fill Phase is used for the control of undesired filamentous organism growth or biological phosphorus release, then the following changes should occur over the time allotted to Static Fill Phase:

TABLE 19.2 Performance Monitoring Table for Static Fill Phase

Parameter	Value Obtained during Static Fill Phase		
	Beginning	Mid-point	End
Alkalinity (mg/L) (as CaCO$_3$)			
pH (standard units)			
Orthophosphate-phosphorus (mg/L)			
ORP (mV)			

TABLE 19.3 Performance Monitoring Table for Mix Fill Phase

Parameter	Value Obtained during Mix Fill Phase		
	Beginning	Mid-point	End
Dissolved oxygen (mg/L)			
Nitrate-nitrogen (mg/L)			
Alkalinity (mg/L) (as CaCO$_3$)			
pH (standard units)			
Orthophosphate-phosphorus (mg/L)			
ORP (mV)			

- Increase in acid production and destruction of alkalinity and drop in pH
- Increase in orthophosphate-phosphorus concentration
- Decrease in ORP value

Table 19.2 lists the parameters to be monitored for the Static Fill Phase and provides entry blocks to record data obtained from testing of collected samples and *in situ* monitoring. The presence of free molecular oxygen or nitrates would hinder the development of an anaerobic/fermentative condition or Static Fill Phase.

MIX FILL PHASE

The Mix Fill Phase is an anaerobic/anoxic condition. If the Mix Fill Phase is used for the control of undesired filamentous organism growth or denitrification, then the following changes should occur over the time allotted to Mix Fill Phase:

- Decrease in residual dissolved oxygen value and eventual loss of dissolved oxygen; the initial dissolved oxygen level should be <1 mg/L in order to promote an oxygen gradient within floc particles
- Decrease in nitrate-nitrogen concentration
- Increase in alkalinity and perhaps an increase in pH
- Decrease in orthophosphate-phosphorus concentration
- Decrease in ORP value

Table 19.3 lists the parameters to be monitored for the Mix Fill Phase and provides entry blocks to record data obtained from testing of collected samples or *in situ* monitoring. The presence of dissolved oxygen ≥1 mg/L would not permit an anaerobic/anoxic condition or Mix Fill Phase.

TABLE 19.4   Performance Monitoring Table for Aerated Fill Phase

Parameter	Value Obtained during Aerated Fill Phase		
	Beginning	Mid-point	End
Dissolved oxygen (mg/L)			
Alkalinity (mg/L) (as CaCO₃)			
pH (standard units)			
Ammonical-nitrogen (mg/L)			
Nitrate-nitrogen (mg/L)			
Orthophosphate-phosphorus (mg/L)			
ORP (mV)			

AERATED FILL PHASE

The Aerated Fill Phase may consist of the entire Fill Phase or the last portion of a Fill Phase that is combined with a Static Fill Phase and/or a Mix Fill Phase. During the Aerated Fill Phase, soluble cBOD is degraded and carbon dioxide is released to the bulk solution. As carbon dioxide dissolves in the bulk solution, carbonic acid is produced and the alkalinity and pH of the bulk solution decrease. Ammonium is removed as a nutrient source for the growth of bacterial cells or sludge (MLVSS), and orthophosphate is also removed as a nutrient source for bacterial growth. If a Static Fill Phase was operated before the Aerated Fill Phase, then biological phosphorus uptake would occur and a relatively large quantity of orthophosphate is removed from the bulk solution. With continued aeration and degradation of organic pollutants (cBOD), the ORP value of the Aerated Fill Phase increases.

If the Aerated Fill Phase is relatively long in time and favorable conditions develop, it is possible for nitrification to occur during this phase. If nitrification occurs, nitrate should be produced. Nitrite production should not occur. Table 19.4 lists the parameters to be monitored for the Aerated Fill Phase and provides entry blocks to record data obtained from testing of collected samples or *in situ* monitoring.

REACT PHASE

React Phase is an aerated condition. Several microbial events occur during the React Phase. Insoluble cBOD is adsorbed by bacterial cells in floc particles, and soluble cBOD is absorbed by the bacterial cells. Soluble cBOD is degraded, resulting in the release of carbon dioxide to the bulk solution and the production of carbonic acid and new bacterial cells (sludge or MLVSS). The production of new cells results in the removal of ammonium and orthophosphate from the bulk solution. If the Static Fill Phase was used previously to the React Phase, then biological phosphorus uptake occurs. This results in a significant decrease in orthophosphate in the bulk solution. With decreasing cBOD, the dissolved oxygen concentration in the SBR increases as well as the ORP value.

If the React Phase is operated to promote nitrification, then the concentration of ammonium decreases in the SBR, while the concentration of nitrate increases.

TABLE 19.5 Performance Monitoring Table for the React Phase

Parameter	Value Obtained during Mix Fill Phase		
	Beginning	Mid-point	End
Dissolved oxygen (mg/L)			
Alkalinity (mg/L) (as CaCO₃)			
pH (standard units)			
Ammonical-nitrogen (mg/L)			
Nitrate-nitrogen (mg/L)			
Nitrite-nitrogen (mg/L)			
Orthophosphate-phosphorus (mg/L)			
ORP (mV)			

TABLE 19.6 Performance Monitoring Table for Carbon Fed Denitrification Period after Settle Phase

Parameter	Value Obtained during Carbon Fed Denitrification Period after Settle Phase		
	Beginning	Mid-point	End
Dissolved oxygen (mg/L)			
Alkalinity (mg/L) (as CaCO₃)			
pH (standard units)			
Nitrate-nitrogen (mg/L)			
ORP (mV)			
Carbon used (quantity)			

Nitrite production should not occur. Nitrification results in a loss of alkalinity and a decrease in pH. Table 19.5 lists the parameters to be monitored for the React Phase and provides entry blocks to record data obtained from testing of collected samples or *in situ* monitoring.

CARBON-FED DENITRIFICATION PERIOD AFTER SETTLE PHASE

If the quantity of nitrate in the decant would result in a permit violation for nitrate or total nitrogen discharge, a soluble carbon source may be added to the SBR after the Settle Phase. The addition of soluble carbon would remove any residual dissolved oxygen and then quickly permit denitrification. The quantity of soluble carbon needed for denitrification to reduce the nitrate concentration to an acceptable level is typically 3 parts carbon to 1 part nitrate. Soluble carbon addition after the Settle Phase would give rise to floating solids that capture gas bubbles released during denitrification. In order to remove the entrapped gases and permit the solids to settle, a short period of aeration or mixing after denitrification would be provided to strip the gases from the solids. Table 19.6 lists the parameters to be monitored for the "carbon-fed denitrification period after the Settle Phase" and provides entry blocks to record data obtained from testing of collected samples or *in situ* monitoring.

20

ORP

Oxidizers are compounds including bromine (Br_2), chlorine (Cl_2), oxygen (O_2), and ozone (O_3) that have the ability to "steal" electrons from other compounds (Figure 20.1). Electrons are negatively charged particles that orbit the nucleus or body of atoms that make the compounds. Therefore, when oxidizers steal electrons, the charge of the oxidizer that gains the electron becomes more negative (for example, +3 to +2 or −1 to −2), while the charge of the compound that loses the electron becomes more positive (for example, +2 to +3 or −2 to −1).

Whenever an electron is transferred from one compound to another, an oxidation–reduction reaction occurs; that is, a change in the net charge of two compounds occurs. One compound becomes more positive and one compound becomes more negative; that is, oxidation–reduction has occurred. When the oxidation–reduction occurs inside living cells, it is called a biochemical reaction. A biochemical reaction occurs each time the substrate (BOD) is degraded.

There are four oxidizers (Table 20.1) that are used by bacteria to degrade substrate and are of concern to operators with respect to the following criteria:

- Type of substrate degraded
- Type and consequence of compounds produced from the degradation of substrate
- Quantity of sludge produced from the degradation of substrate
- Cost to provide the oxidizer

The four oxidizers used by bacteria to degrade substrate include free molecular oxygen, nitrate, sulfate, and an organic molecule. There are many organic molecules that can be used as oxidizers, and collectively they are represented by the chemical

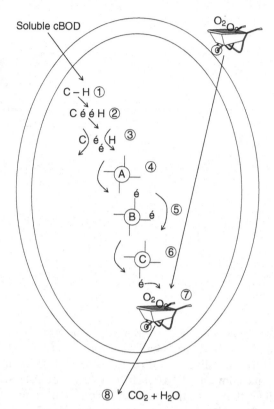

Figure 20.1 *Transfer of electrons. Soluble cBOD consists of organic molecules that possess carbon–hydrogen bonds (**1**) that are made of two electrons (**2**). When soluble cBOD enters the bacterial cell and is degraded, the carbon–hydrogen bonds are broken by cellular enzymes (**3**). The freed electrons are transferred to an electron transport molecule (A, B, and C). As the electrons travel down the transport molecule, some of the energy in the electrons is released to the cell (**4, 5,** and **6**). The release energy is stored in high-energy phosphate bonds inside the cell. Eventually, the electrons must be removed from the transport molecule. This is achieved by "dumping" the electrons on a final electron transport molecule or wheelbarrow such as free molecule oxygen (**7**), nitrate, sulfate, or another organic molecule. The wheelbarrow transports the electrons from the cell and dumps them along with associated wastes or products to the bulk solution.*

TABLE 20.1   Oxidizers of Concern to Wastewater Treatment Plant Operators

Oxidizer	Formula
Free molecular oxygen	O_2
Nitrate	NO_3^-
Sulfate	SO_4^{2-}
Organic molecule	CH_2O^a

[a]Simplistic formula to denote any organic molecule.

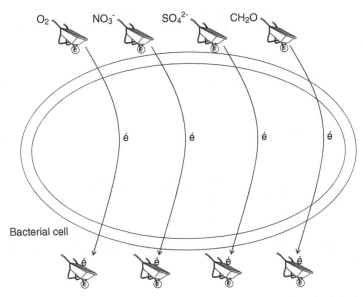

Figure 20.2 *Electron transport molecules or wheelbarrows. There are four final electron transport molecules or wheelbarrows that are of concern to wastewater treatment plant operators. These wheelbarrows are free molecular oxygen (O_2), nitrate (NO_3^-), organic molecules (CH_2O), and sulfate (SO_4^{2-}). These molecules are used to capture freed electrons from broken chemical bonds in BOD and transport the electrons from the cell to the bulk solution. A different variety of wastes are produced from each wheelbarrow used. Therefore, it is important for an operator to know what wheelbarrow is being used and where (sewer system or treatment tank) that wheelbarrow is being used. All four wheelbarrows can be used to degrade cBOD, but only the free molecular oxygen wheelbarrow can be used to degrade nBOD, that is, nitrify.*

TABLE 20.2   Hierarchy of Electron Transport Molecules or Wheelbarrows

Quantity of cBOD as Acetate Degraded	Wheelbarrow Used for Acetate Degradation	Approximate Weight of Offspring (Sludge) Produced
1	O_2	0.6 pound
1	NO_3^-	0.4 pound
1	SO_4^{2-}	0.04–0.1 pound
1	CH_2O	0.04–0.1 pound

formula CH_2O. These oxidizers steal electrons as chemical bonds in substrate are broken, and electrons are released inside the bacterial cell. The oxidizers "carry" or transport the electrons from inside the cell to outside the cell; that is, the oxidizers serve as the final electron transport molecules or "wheelbarrows" (Figure 20.2).

The degradation of substrate occurs inside the bacterial cell. Therefore, the bacterial cell absorbs substrate or soluble cBOD and oxidizers. However, bacterial cells can only use one wheelbarrow at any time regardless of the types of wheelbarrows that are available for use. Also, there is a "pecking" order or hierarchy with selection of wheelbarrows that are used at any time to degrade the substrate. This hierarchy is based upon the wheelbarrow that provides the most energy and offspring (new bacterial cells or sludge) (Table 20.2).

The use of wheelbarrows to perform oxidation–reduction reactions to degrade soluble cBOD in any wastewater treatment facility is based upon the following conditions:

- Bacteria present
- Wheelbarrows present
- Wheelbarrow gradients
- Substrates present
- Oxidation–reduction potential

Although all conditions are important for an oxidation–reduction reaction to occur, it is the measurement of the oxidation–reduction potential when supported with appropriate testing that permits an operator to determine (a) the bacterial activity that is occurring, (b) its consequences to the treatment facility, and (c) whether operational conditions should be changed to future promote the activity or terminate the activity.

BACTERIA PRESENT

No bacterium can use all electron carrier molecules or wheelbarrows; that is, no bacterium has all the necessary "tools" or enzymes for using all wheelbarrows. Also, no wheelbarrow can be used by all bacteria. Some bacteria such as strict aerobes can only use one wheelbarrow, molecular oxygen. Some bacteria such as facultative anaerobes can use a variety of wheelbarrows, free molecular oxygen, nitrate, and an organic molecule.

Bacteria are aerobic, facultative anaerobic, or anaerobic with respect to their response to free molecular oxygen (Figure 20.3). Examples of aerobic bacteria

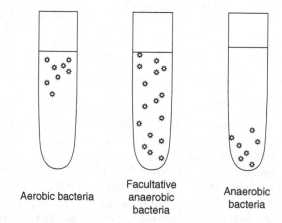

Aerobic bacteria   Facultative anaerobic bacteria   Anaerobic bacteria

Figurer 20.3 Growth of bacteria in the presence of free molecular oxygen. In test tubes that contain a relatively low concentration of dissolved oxygen, aerobic bacteria would be found near the top of the tube where oxygen is greatest and facultative anaerobic bacteria would be found near the top of the tube, middle of the tube where oxygen is minimal, and bottom of the tube where no oxygen exists. Anaerobic bacteria would be found at the bottom of the tube where no oxygen exists.

include *Nitrosomonas* and *Zoogloea*. They use only free molecular oxygen to degrade nBOD and cBOD, respectively. *Nitrosomonas* is a nitrifying bacterium. *Zoogloea* is a floc-forming bacterium.

Examples of facultative anaerobic bacteria include *Bacillus* and *Pseudomonas*. They use free molecular oxygen to degrade cBOD and nitrate in the absence of free molecular oxygen to degrade cBOD. When nitrate is used by these bacteria, they are referred to as denitrifying bacteria. *Escherichia coli* also are facultative anaerobic bacteria that are capable of using free molecular oxygen, nitrate, and an organic molecule to degrade soluble cBOD. *Escherichia coli* are floc-forming bacteria and denitrifying bacteria. When *Escherichia coli* use an organic molecule to degrade soluble cBOD, a mixture of acids and alcohols is produced. Some of the acids are volatile and malodorous.

Anaerobic bacteria either are inactive or die in the presence of free molecular oxygen. Sulfate-reducing bacteria (SRB) such as *Desulfovibrio* use sulfate to degrade a relatively small number of substrates or organic molecules. The use of sulfate is referred to as sulfate reduction. During sulfate reduction, hydrogen sulfide and sulfide are produced. Sulfate-reducing bacteria are inactive in the presence of free molecular oxygen. Methane-forming bacteria also are anaerobes. Methane-forming bacteria die in the presence of free molecular oxygen at 0.01 mg/L.

WHEELBARROWS PRESENT

All four wheelbarrows used in the oxidation–reduction or degradation of soluble cBOD are found in relatively large quantities in biological wastewater treatment facilities that nitrify (Table 20.3). Free molecular oxygen is provided through aeration of the mixed liquor. Although some nitrate may be discharged to the sewer system by an industry, nitrate is produced in the mixed liquor through nitrification. Sulfate is found in the influent of domestic and municipal wastewater treatment plants and may be produced in the mixed liquor through aeration and sulfide oxidation. Sulfate is a component of urine that is discharged to the sewer system. Sulfate also is a component of groundwater that enters a sewer system through inflow and infiltration (I/I).

Sulfate is produced through the biological and chemical oxidation of sulfide. Sulfide is formed under anaerobic conditions by sulfate-reducing bacteria in sewer systems, upset thickeners, and anaerobic digesters. Sulfide also is produced through the anaerobic degradation of sulfur-containing amino acids cysteine and methionine and proteins that contain these amino acids (Figure 20.4). Cysteine contains the thiol group (-SH), while methionine contains sulfur as "S." When the thiol group and sulfur are released during the degradation of the amino acids, sulfide is formed. Sulfur-oxidizing bacteria such as *Thiobacillus* oxidize sulfide to sulfate in the aerated mixed liquor.

Organic compounds (CH_2O) are present in relatively large quantities in domestic and municipal wastewaters. These compounds are present in colloidal, particulate, and soluble forms. Industrial discharges usually contain many soluble organic compounds and contribute to elevated levels of soluble cBOD in municipal wastewater treatment plants.

TABLE 20.3   Sources of Wheelbarrows

Wheelbarrow	Source
Free molecular oxygen (O_2)	Aeration of mixed liquor
Nitrate (NO_3^-)	Nitrification in the mixed liquor
Sulfate (SO_4^{2-})	Urine and groundwater (I/I)
Organic compound (CH_2O)	Influent cBOD

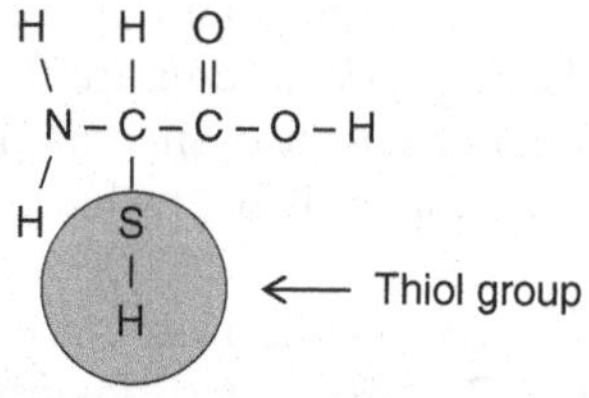

Cysteine   ($NH_2CHSHCOOH$)

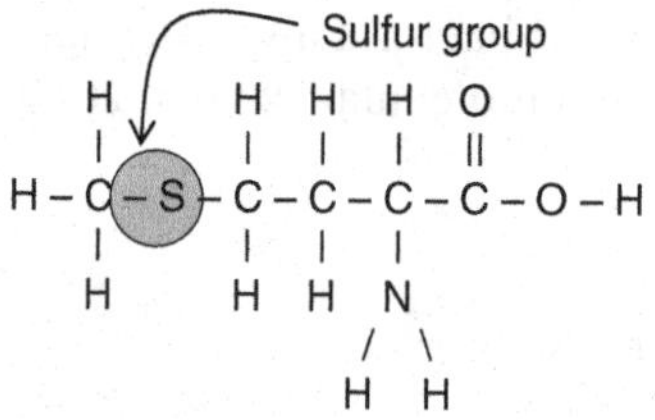

Methionine   ($CH_3SCH_2CH_2NH_2CHCOOH$)

Figure 20.4 *Cysteine and methionine. There are two sulfur-containing amino acids. These amino acids are cysteine ($NH_2CSHCHCOOH$) and methionine ($CH_3SCH_2CH_2NH_2CHCOOH$). Cysteine contains sulfur in a thiol group (-SH), while methionine contains sulfur (S) as part of the aliphatic chain.*

Ammonium (NH_4^+) is the major soluble nBOD in wastewater and serves as a nitrogen nutrient for bacteria and an energy substrate for nitrifying bacteria. Ammonium is released from urea (H_2NCONH_2) that is found in urine. Organic-nitrogen compounds such as amino acids and proteins that are found in fecal waste contain amino groups (-NH_2) that produce ammonium in the mixed liquor when they are degraded. Additional organic-nitrogen compounds include surfactants and polymers.

WHEELBARROW GRADIENTS

Several wheelbarrows may be used at the same time by different bacteria in the sewer system or treatment tank, if the thickness of the biofilm (Figure 20.5) or floc particle (Figure 20.6) provides for a wheelbarrow gradient. If the thickness of the biofilm is >50 μm or the diameter of the floc particle is >100 μm and the dissolved oxygen in the bulk solution is <1 mg/L and nitrate and soluble cBOD are available, then degradation of soluble cBOD can occur with the use of free molecular oxygen and nitrate. Depending on the thickness of the biofilm or the diameter of the floc

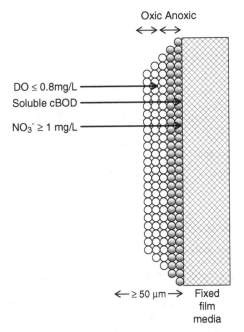

Figure 20.5 *Wheelbarrow gradient across biofilm. A wheelbarrow gradient for dissolved oxygen (DO) and nitrate (NO_3^-) can be established across biofilm, if (1) the DO in the bulk solution is ≤0.8 mg/L, (2) the nitrate in the bulk solution is ≥1 mg/L, (3) there is adequate soluble cBOD to penetrate to the fixed film media, and (4) the biofilm is ≥50 μm in thickness. Under this condition the DO would not penetrate to the bottom of the biofilm. Only the outer layer of bacteria (clear spheres) would have access to free molecular oxygen to degrade the soluble cBOD. This is the oxic layer. The layer of bacteria beneath the oxic layer is the anoxic layer. Here, the bacteria (shaded spheres) would only have nitrate available for use to degrade the soluble cBOD.*

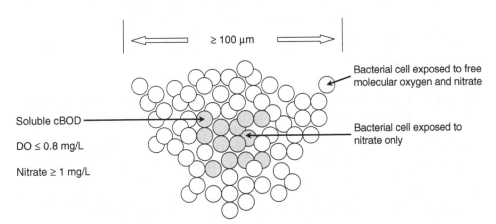

Figure 20.6 *Wheelbarrow gradient across floc particle. A wheelbarrow gradient for dissolved oxygen (DO) and nitrate (NO_3^-) can be established across a floc particle, if (1) the DO in the bulk solution is ≤0.8 mg/L, (2) the nitrate in the bulk solution is ≥1 mg/L, (3) there is adequate soluble cBOD to penetrate to the fixed film media, and (4) the floc particle is ≥100 μm in diameter. Under this condition the DO would not penetrate to the core of the floc particle. Only the bacteria on the perimeter of the floc particle (clear spheres) would have access to free molecular oxygen to degrade the soluble cBOD. This is the oxic perimeter. The bacteria in the core of the core of the floc particle would only have access to nitrate. This is the anoxic core. Here, the bacteria (shaded spheres) would use nitrate to degrade the soluble cBOD.*

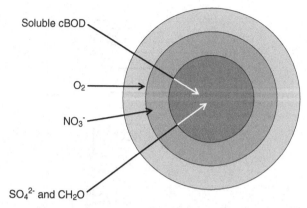

Soluble cBOD

O_2

NO_3^-

SO_4^{2-} and CH_2O

Figure 20.7 *Use of multiple wheelbarrows. If residual quantities of free molecular oxygen, nitrate, and sulfate and sufficient soluble cBOD or soluble organic molecules (CH_2O) are present in the bulk solution and the floc particle is medium (150–500 µm) or large (>500 µm) in size, then the use of several final electron carrier molecules or wheelbarrows can occur simultaneously. As the soluble cBOD or CH_2O penetrates to the core of the floc particle from the perimeter of the floc particle, the bacteria on the perimeter or outer circle will use free molecular oxygen to degrade the soluble cBOD. The next circle of bacteria will have soluble cBOD to degrade but lack free molecular oxygen due to its depletion in the outer circle of bacteria. The bacteria in the next circle from the perimeter of the bacteria will use nitrate to degrade the soluble cBOD. In the inner circle of bacteria where free molecule oxygen and nitrate are not available, sulfate-reducing bacteria (SRB) and fermentative bacteria or acid-forming bacteria will use sulfate and CH_2O, respectively, to degrade soluble cBOD. Under this operational condition, it is possible for four wheelbarrows to be used at the same time.*

particle and types of wheelbarrows available, several wheelbarrow gradients can occur and the degradation of soluble cBOD can occur with several wheelbarrows at the same time in the same flocculated mass of bacteria (Figure 20.7).

SUBSTRATE PRESENT

The degradation of cBOD can occur with the use of free molecular oxygen, nitrate, sulfate, and an organic molecule. The degradation of nBOD or nitrogenous BOD can only occur with the use of free molecular oxygen.

cBOD can be found in the sewer system and treatment tanks as soluble, colloidal, and particulate substrate. In the soluble form, cBOD is quickly absorbed by bacterial cells as it passes through the cell wall and cell membrane. Inside the cell it is degraded (Figure 20.8). Therefore, the degradation of soluble cBOD exerts an immediate demand for wheelbarrows or final electron transport molecules.

Colloidal and particulate cBOD cannot pass through the cell wall and cell membrane. These substrates are adsorbed by the polysaccharide coating that surrounds many of the bacterial cells in the biofilm or floc particle (Figure 20.9). Over time, the polysaccharide-coated cells produce exoenzymes. The exoenzymes pass through the cell membrane and cell wall to the polysaccharide coating where they come in contact with the adsorbed substrate and split the colloidal and particulate cBOD into simplistic soluble substrates that are then easily absorbed by the bacterial cells. Here, the soluble cBOD is degraded within the bacteria by endoenzymes. Therefore,

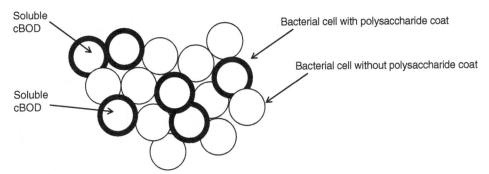

Figure 20.8 *Absorption and degradation of soluble cBOD. Simplistic soluble cBOD can be absorbed directly by bacteria with and without a polysaccharide coat; that is, they diffuse quickly through the cell wall and the cell membrane of the bacterial cell. These forms of cBOD consist of simple sugars (monosaccharides), simple alcohols, simple organic acids, simple amino acids, and simple lipids. These forms need not be adsorbed to a polysaccharide coat and hydrolyzed—that is, split into small molecules for absorption by the bacterial cell.*

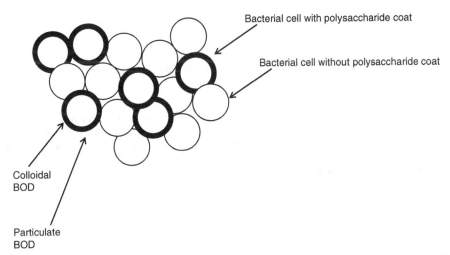

Figure 20.9 *Absorption of colloidal and particulate BOD. Colloidal BOD and particulate BOD is removed from the waste stream by bacteria that have a polysaccharide coat. Colloidal BOD and particulate BOD are adsorbed (come in contact) by the polysaccharide coat. If sufficient retention time is available, the adsorbed BOD is solubilized and degraded (stabilization). This form of removal and degradation of BOD is similar to the contact-stabilization mode of operation used at many activated sludge processes.*

the degradation of colloidal and particulate cBOD does not exert an immediate demand for wheelbarrows or final electron transport molecules. This mode of degradation of colloidal and particulate cBOD where substrate is first adsorbed (in contact with the bacterial cell) and then solubilized and degraded (stabilize) is similar to the "contact-stabilization" mode of operation that is used at many activated sludge process (Figure 20.10).

Nitrogenous BOD or nBOD consists of two nitrogen-containing compounds, ammonium (NH_4^+) and nitrite (NO_2^-). Ammonium is the energy substrate for

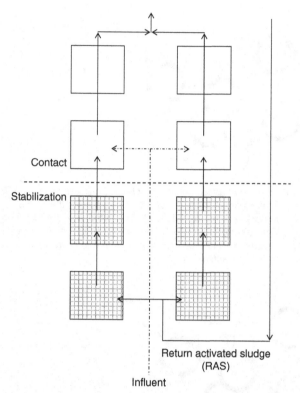

Contact

Stabilization

Return activated sludge
(RAS)

Influent

Figure 20.10 *Contact-stabilization mode of operation. In the contact-stabilization mode of operation, return activated sludge (RAS) that contains bacteria with adsorbed substrate enters the upstream (shaded) aeration tanks, where long retention time is provided for the degradation (stabilization) of colloidal BOD and particulate BOD (substrate). Influent enters the downstream (clear) aeration tanks, where colloidal BOD and particulate BOD come in contact with floc particles that have bacterial cells with a polysaccharide coating that adsorb or remove these BOD from the waste stream.*

ammonia-oxidizing bacteria (AOB), including *Nitrosomonas* and *Nitrosospira*, whereas nitrite serves as the energy substrate for nitrite-oxidizing bacteria (NOB), including *Nitrobacter* and *Nitrospira*. Ammonium is found in relatively large quantities in the influent and is released in the mixed liquor when organic-nitrogen compounds such as amino acids and proteins are degraded. Unless nitrite is discharged to the sewer system by an industry, it should not be found in the influent. Nitrite is produced in the mixed liquor through nitrification. Typically, nitrification does not result in the accumulation of nitrite.

OXIDATION–REDUCTION POTENTIAL

Potential is the "ability" of a biochemical reaction (degradation of substrate and use of a wheelbarrow) to occur; in electrical terms, oxidation–reduction potential (redox), or ORP, is the potential energy as measured in volts. With respect to ORP, the potential or voltage is relatively small and is measured in millivolts when a metal

TABLE 20.4 ORP (Range in Values) and Biochemical Reactions

Biochemical Reaction	Wheelbarrow Used	ORP (mV)
Nitrification	O_2	+100 to +350
cBOD, oxidation	O_2	+50 to +250
Biological phosphorus uptake	O_2/NO_3^-	+25 to +250
Denitrification	NO_3^-	+50 to –50
Sulfate reduction	SO_4^{2-}	–50 to –250
Biological phosphorus release	CH_2O	–100 to –250
Acid formation	CH_2O	–100 to –225
Methane production	CO_2/CH_2O	–175 to –400

probe is placed in an aqueous solution (water, wastewater, or sludge) in the presence of an oxidizing agent (O_2, NO_3^-, SO_4^{2-}, or CH_2O) and reducing agent such as carbonaceous (cBOD) and ammonium (polluting wastes). The measured millivolts provide an indicator of the ability of the oxidizers in the aqueous solution to perform specific biochemical reactions to reduce the cBOD or ammonium in the aqueous solution.

Commonly used *guideline ranges* of ORP values to indicate the biochemical reactions occurring and the oxidizing agents or wheelbarrows that are used are presented in Table 20.4. The biochemical reactions in the table are listed from top to bottom according to most positive millivolt values to most negative millivolt values. The use of these values for ORP monitoring should be correlated appropriately with field and laboratory data obtained in the sewer system and treatment tanks for (1) alkalinity and pH, (2) ammonium, nitrite, and nitrate, (3) dissolved oxygen, (4) sulfate and sulfide, and (5) orthophosphate values, depending upon the biochemical reaction desired or detected such as nitrification. The ORP range of values for nitrification should be correlated to (1) production of nitrite and/or nitrate, (2) decrease in alkalinity and pH, and (3) dissolved oxygen value.

+350 TO +100mV, NITRIFICATION

Within this millivolt range of values, ammonium is oxidized to nitrite, and nitrite is oxidized to nitrate by nitrifying bacteria with the use of free molecular nitrogen. Nitrification accelerates as cBOD is degraded and ORP increases. Nitrification results in the loss of alkalinity and decrease in pH. Dissolved oxygen is used as the final electron transport molecule or wheelbarrow.

+250 TO +50mV, cBOD DEGRADATION WITH FREE MOLECULE OXYGEN

Within this millivolt range of values, cBOD is degraded by organotrophic bacteria with the use of free molecular oxygen. The BOD is degraded to nonpolluting wastes and less polluting wastes. The quantity of cBOD decreases as ORP increases. Carbonic acid is formed during the degradation of cBOD, and alkalinity and pH decrease as a consequence of the production of carbonic acid. Dissolved oxygen is used as the final electron transport molecule or wheelbarrow.

+250 TO +25 mV, BIOLOGICAL PHOSPHORUS UPTAKE

Within this millivolt range of values following an anaerobic/fermentative condition, uptake of orthophosphate by poly-P bacteria occurs. There are two final electron carrier molecules or wheelbarrows that are used over this ORP range of values: free molecular oxygen and nitrate. During biological phosphorus uptake the concentration of orthophosphate in the bulk solution decreases with increasing ORP.

+50 TO –50 mV, DENITRIFICATION

Within this millivolt range of values, cBOD is degraded to nonpolluting wastes and less polluting wastes by facultative anaerobic (denitrifying) bacteria with the use of nitrate. This is denitrification. During denitrification the concentration of nitrate decreases, alkalinity increases and pH increase with decreasing ORP. Insoluble molecular nitrogen, nitrous oxide, and carbon dioxide are produced during denitrification, and these gases are responsible for "clumping" or "dark sludge rising." Nitrate is used as the final electron transport molecule or wheelbarrow.

–50 TO –250 mV, SULFATE REDUCTION

Within this millivolt range of values, a limited number of substrates or soluble cBOD are degraded to nonpolluting wastes and less polluting wastes by sulfate-reducing bacteria. This is sulfate reduction. During sulfate reduction the concentration of sulfate decreases, the concentration of sulfide/hydrogen sulfide increases, and alkalinity and pH increase with decreasing ORP. Sulfate is used as the final electron transport molecule or wheelbarrow. In addition to the production of sulfide/hydrogen sulfide through sulfate reduction, sulfide is released to the bulk solution during anaerobic degradation of the sulfur-containing amino acids, cysteine and methionine.

–100 TO –250 mV, BIOLOGICAL PHOSPHORUS RELEASE

Within this millivolt range of values, poly-P bacteria release orthophosphate to the bulk solution. Under this anaerobic/fermentative condition, facultative anaerobic or acid-forming bacteria produce fatty acids that "trigger" the release of orthophosphate from poly-P bacteria. Biological phosphorus release results in an increase in the concentration of orthophosphate in the bulk solution and decreases in pH and alkalinity occur as the ORP value decreases.

ORP MONITORING

An ORP probe is a millivolt meter. It measures the voltage across a circuit formed by a negative pole or reference electrode that is identical to the pH reference electrode and is made of silver and a positive pole or measuring electrode that is made

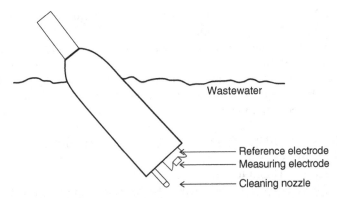

Figure 20.11 *ORP probe. The ORP (oxidation–reduction potential) probe or redox probe is used for measuring the millivolt (mV) of an aqueous sample of water, wastewater, or sludge. The reading provides an indicator with respect to the type of biological activity (such as nitrification and denitrification) that may occur in the sample. Although there are two electrodes in an ORP probe, the reference electrode and the measuring electrode, the design and construction of the probe gives the appearance of only one electrode present in the probe. Some ORP probes have a cleaning nozzle to keep debris and scum from the surfaces of the reference electrode and measuring electrode.*

of a noble metal such as platinum, gold or silver. The surface of the measuring electrode accumulates charge without reacting chemically. This charge is measured relative to the solution, so the solution "ground" voltage comes from the reference junction.

Most ORP probes have both electrodes housed in one body and are referred to as combined electrodes. The construction of the probe gives the impression of just "one" electrode (Figure 20.11). Once immersed in an aqueous solution, the probe measures the voltage across the circuit formed by these the two electrodes. Voltage measurements are expressed as millivolts and may be positive or negative. A positive voltage indicates an aqueous solution that attracts electrons (oxidizing condition), whereas a negative voltage indicates an aqueous solution that loses electrons (reducing condition). Therefore, chlorinate water containing hypochlorous ions (OCl^-) displays a positive ORP value, while a wastewater containing formate ($HCOOH$), oxalic acid ($C_2H_4O_4$), hydrogen sulfide (H_2S), or hydrogen sulfite (HSO_3^-) displays a negative ORP value.

The reference electrode is surrounded by an electrolyte (salt solution) that produces a constant, steady, and relatively small voltage. The electrode serves as a "reference" against which the voltage generated by the measuring electrode and oxidizers in the aqueous solution may be compared. The meter monitors the difference in voltage between the reference electrode and measuring electrode.

Initial studies of ORP electrodes at Harvard University in 1936 indicated a strong correlation between ORP and bacterial activity. Because a direct correlation exists between ORP and inactivation of *Escherichia coli* and several pathogens in "clean" water, it is possible to select an ORP value at which a predictable level of disinfection can be achieved regardless of changes in quality and quantity of reducing agents or pollutants that are present.

The World Health Organization (WHO) adopted an ORP value of +700 mV as a standard for safe drinking water. The German Standards Agency accepted an ORP value of +750 mV for public pools, and the National Swimming Institute adopted an ORP value of +650 mV for public spas.

ORP is an acceptable and convenient measure of the ability (potential) of oxidizers or reducers (oxidation–reduction) to perform chemical reactions. However, ORP is not valid over a wide range of pH values. The higher the pH in clean water, the lower the ORP value. This is due to the absence of antioxidant capacity or reducing agents (pollutants) in the water. Also, ORP is not corrected for temperature.

As with all testing, ORP has limitations. Although ORP can be used as an indicator of bacterial activity in wastewater treatment processes, it cannot be used directly due to the effects of pH and temperature. Therefore, general ranges of ORP values are provided for when specific biochemical reactions in polluted water or wastewater occur. However, ORP can be better correlated to treatment processes by checking the presence of oxidizers (dissolved oxygen, nitrate, and sulfate) and reducers (cBOD, COD, ammonium, and sulfides) with wet tests and measurements of changes in alkalinity and pH. The best recommendation for ORP is to use wet tests and correlate the ORP values to test values.

The oxidation–reduction potential system can be used to (1) determine the health of a treatment process and (2) establish the most favorable conditions for specific bacterial activity. For example, with respect to the health of a treatment process, under an anoxic condition, if nitrate production has been inhibited or little nitrate production has occurred, the ORP value of the treatment process would not be able to indicate an ORP value that would be indicative of denitrification.

Examples that illustrate the establishment of a favorable condition for specific bacterial activities include:

- Confirmation of the presence of oxic (aerobic), anoxic, or anaerobic/fermentative conditions and the expected waste products for that condition.
- Development of the most favorable operational conditions for biological activities such as nitrogen or phosphorus removal. Different operation conditions are required for nitrogen removal and phosphorus removal.

The operation of the SBR can be monitored and controlled and its performance improved with (1) knowledge of bacterial activity under oxic, anoxic, and anaerobic/fermentative conditions and (2) use of ORP that is supported by wet tests and alkalinity and pH testing.

ORP monitoring may be performed as *in situ* or in the laboratory and may be tested continuously or as needed. Monitoring may be performed by field and laboratory probes. Regardless of the frequency of monitoring and type of probe that is used, the probe should be placed approximately 2 ft below the surface of the wastewater during field testing. Also, critical factors that should be addressed when purchasing an ORP probe include:

- Accuracy
- Frequency of calibration
- Frequency of cleaning
- Stability or drifting of value

In dilute aqueous solutions of oxidizers and reducers, it will take time to accumulate a measurable charge. Also, the ability of an aqueous solution to resist change in the presence of strongly oxidizing or strongly reducing compounds is a measure of the system "poise." An aqueous solution with little poise will experience significant changes in ORP with the addition of a strong oxidizer such as hypochlorous acid (HOCl) or strong reducer such as a slug discharge of methanol (CH$_3$OH).

21

Microscopy

The use of the microscope to support process control and troubleshooting efforts in the activated sludge process is extensively reviewed in *Microscopic Examination of the Activated Sludge Process*. There are several microscopic components of the mixed liquor that serve as indicators or "bioindicators" of the "health" of the biomass and the expected quality of the decant. These parameters include the following:

- Bulk solution
- Floc particles
- Filamentous organisms
- Ciliated protozoa
- Metazoa

BULK SOLUTION

In healthy, mature, and stable mixed liquor the bulk solution would contain insignificant dispersed growth (Figure 21.1). In unhealthy mixed liquor the bulk solution would contain significant dispersed growth (Figure 21.2) or excessive dispersed growth (Figure 21.3).

FLOC PARTICLES

In healthy, mature, and stable mixed liquor the floc particles would be mostly medium (150–500 μm) and/or large (>500 μm) in size. Due to the presence of an

Troubleshooting the Sequencing Batch Reactor, by Michael H. Gerardi
Copyright © 2010 by John Wiley & Sons, Inc.

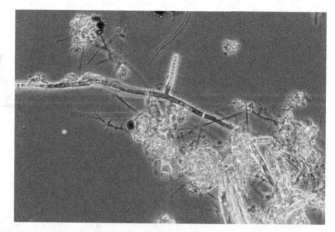

Figure 21.1   *Insignificant dispersed growth. Dispersed growth consists of small (<10μm) and spherical floc particles. In a "healthy" biomass the relative abundance of dispersed growth is "insignificant" or <20 small floc particles or "cells" per field of view at 100× total magnification. The field of view is the circular area observed during a microscopic examination. Because a photomicrograph of the field of view is taken, only a rectangular portion of the field of view is shown. Here, the bulk solution around the floc particle and branched filamentous fungi contains an "insignificant" rating for dispersed growth.*

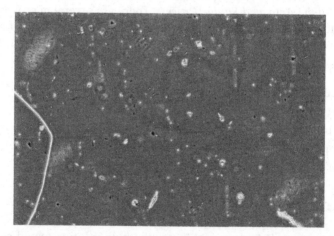

Figure 21.2   *Significant dispersed growth. In an "unhealthy" biomass the relative abundance of dispersed growth may be "significant" or "tens" (20, 30, 40, 50, 60, 70, 80, 90) of small floc particles or "cells" per field of view at 100× total magnification.*

acceptable level of filamentous organism growth, the floc particles would be irregular in shape.

The floc particles would be mostly golden-brown in the core and white on the perimeter. The difference in color is due to the accumulation of oils secreted by old bacteria in the core of the floc particles. The floc particles also would be firm in structure as revealed through methylene blue staining (Figure 21.4). The floc particles would contain little or no amorphous Zoogloeal growth (Figure 21.5) or

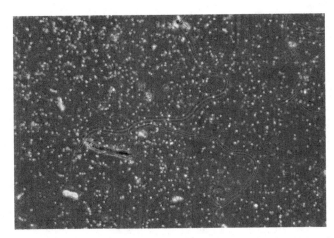

Figure 21.3 *Excessive dispersed growth. In an "unhealthy" biomass the relative abundance of dispersed growth may be "excessive" or "hundreds" (100, 200, 300, ...) of small floc particles or "cells" per field of view at 100× total magnification.*

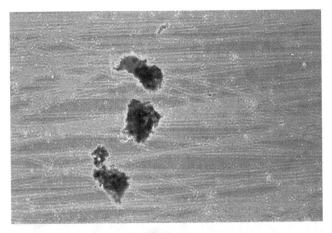

Figure 21.4 *Firm floc particles under methylene blue. The relative strength or density of floc particles can be observed by staining a wet mount of mixed liquor with a drop of methylene blue. Under methylene blue, bacterial cells stain dark blue, whereas biological secretions and absorbed chemicals such as fats, oils, and grease stain light blue. In the three floc particles in this photomicrograph, the particles are dark blue in appearance with no openings or voids in the floc particle. Only on the perimeter of the floc particles is a light blue edge observed. This is due to the production of large quantities of polysaccharides by the rapidly growing bacteria on the edge of the floc particles. The polysaccharides stain light blue. Because most of the area of each floc particle is dark blue, the floc particles appear to be firm and dense.*

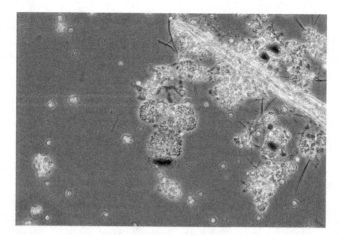

Figure 21.5 *Amorphous Zoogloeal growth. Zoogloeal growth is the rapid and undesired growth of floc-forming bacteria. This growth results in the production of weak and buoyant floc particles due to the separation of the bacteria by a copious quantity of gelatinous material. Amorphous Zoogloeal growth (growth without specific form or shape) is most commonly observed in activated sludge process and can be seen in the center of the photomicrograph as poorly compacted or loosely aggregated, globular masses of bacterial cells.*

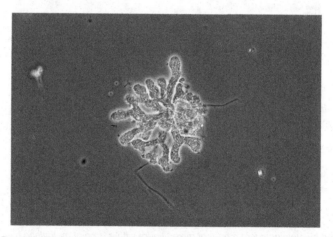

Figure 21.6 *Dendritic Zoogloeal growth. Dendritic ("tooth-like" or "finger-like") Zoogloeal growth is less often observed in activated sludge processes as amorphous Zoogloeal growth. Dendritic usually appears at young sludge ages or mean cell residence times (MCRT).*

dendritic Zoogloeal growth (Figure 21.6), and the floc particles would test negatively to the India ink reverse stain (Figure 21.7).

FILAMENTOUS ORGANISMS

Filamentous organisms perform positive roles in the treatment process. They degrade cBOD and more importantly they provide strength to the floc particle that enables the floc particle to resist turbulence or shearing action and increase in size. However,

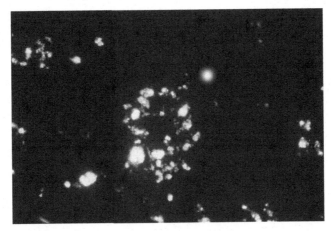

Figure 21.7 *Negative India ink reverse stain. Under an India ink reverse stain using a phase contrast microscope, bacterial cells stain black and/or golden-brown. Stored food or polysaccharides block the penetration of ink (carbon black particles) into the floc particle and therefore appears white. If the majority of the area of the floc particle is black and/or golden-brown rather than white, there is relatively little stored food in the floc particle and a low probability of a nutrient deficiency during the React Phase in an SBR. This is a negative India ink reverse stain.*

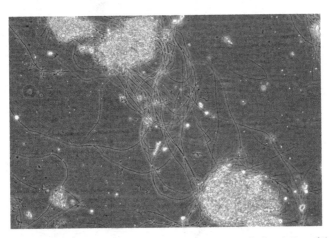

Figure 21.8 *Interfloc bridging. Interfloc bridging is the joining in the bulk solution of the extended filamentous organisms from the perimeter of two or more floc particles. Interfloc bridging adversely affects desired settleability.*

filamentous organisms also perform negative roles in the treatment process. In undesired numbers the filamentous organisms contribute to settleability problems and loss of solids in the decant. Some of the organisms also produce viscous chocolate-brown foam that contributes to numerous operational problems. Therefore, a proper level of filamentous organisms is needed in order to obtain the positive roles and prevent the negative roles. This level would correspond to approximately one to five filamentous organisms in most floc particles.

Also, insignificant interfloc bridging (Figure 21.8) and insignificant open floc formation (Figure 21.9) would be present. These two forms of filamentous organism/floc structure adversely affect solids settleability.

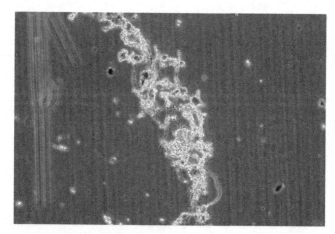

Figure 21.9 *Open floc formation. Open floc formation or diffused floc formation is the scattering of the floc bacteria in many small groups along the lengths of the filamentous organisms in the floc particle. Open floc formation adversely affects desired settleability.*

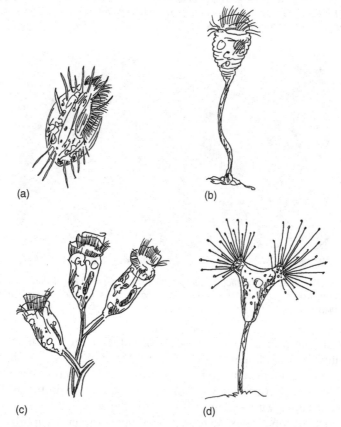

(a) (b)

(c) (d)

Figure 21.10 *Ciliated protozoa associated with high BOD removal efficiency. There are several ciliated protozoa when present in large numbers that serve as bioindicators of a "healthy" biomass that is capable of high BOD removal efficiency. These protozoa consist of the crawling ciliate* Euplotes affinis *(a) and the stalk or tentacle ciliates* Vorticella aquadulcis *(b),* Zoothamnium procerius *(c), and* Acineta tuberose *(d).*

CILIATED PROTOZOA

There are three groups of ciliated protozoa that are commonly found in activated sludge process. These groups include free-swimming ciliates, crawling ciliates, and stalk ciliates. Generally, when present as the dominant protozoan groups in mixed liquor, they are indicative of healthy mixed liquor and an acceptable decant. However, there are several groups of ciliated protozoa that proliferate under adverse operational conditions. Therefore, if ciliated protozoa are used as bioindicators of healthy mixed liquor, attention should be given to the identification of specific ciliated protozoa as indicators of healthy mixed liquor. Ciliated protozoa that serve as bioindicators of high BOD removal efficiency include *Acineta tuberose, Euplotes affinis, Vorticella aquadulcis*, and *Zoothamnium procerius* (Figure 21.10). Ciliated protozoa that are associated with optimal conditions for nitrification include *Blepharisma americanum, Chilodonella minuta, Coleps hirtus*, and *Epistylis plicatilis* (Figure 21.11).

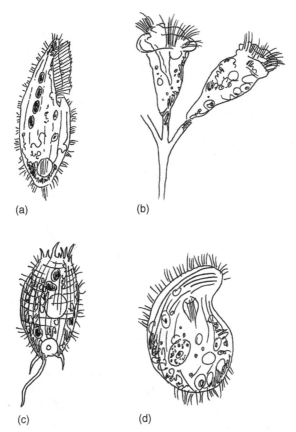

(a)   (b)

(c)   (d)

Figure 21.11 *Ciliated protozoa associated with optimal conditions for nitrification. There are several ciliated protozoa when present in large numbers that serve as bioindicators of a "healthy" biomass that is reflective of optimal conditions for nitrification. These protozoa consist of the free-swimming ciliate* Blepharisma americanum *(a), the stalk ciliate* Epistylis plicatilis *(b) and the free-swimming ciliates* Coleps hirtus *(c) and* Chilodonella minuta *(d).*

METAZOA

The two metazoa of concern as indicators of mixed liquor health and acceptable decant quality are the rotifer (Figure 21.12) and free-living nematode (Figure 21.13). These organisms typically are found in relatively small numbers in activated sludge processes. They are strict aerobes and are sensitive to sudden or significant changes

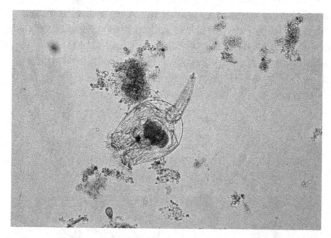

Figure 21.12   *Rotifer. Rotifers are strict aerobes that enter the treatment process through I/I. There are crawling rotifers and free-swimming rotifers such as* Brachinus. *This rotifer has a transparent protective covering or lorica, and beneath the lorica can be seen the digestive system and gonads. From the posterior portion of the rotifer, one can see the foot and toes. The lorica, foot, and toes are calcified.*

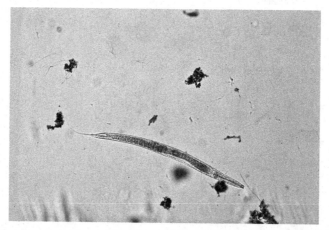

Figure 21.13   *Free-living nematode. The most commonly observed worm in the activated sludge process is the free-living nematode. They enter the treatment system through I/I as soil and water organisms and are strict aerobes. Free-living nematodes do not cause disease and consist of three "tubes." The outer tube is the cuticle. The middle tube is a longitudinal muscular system, and the inner tube is the digestive tract. If the digestive tract contains food, it appears brown. When the cuticle is dispersed by cell bursting agents or surfactants, the digestive tract cannot be seen.*

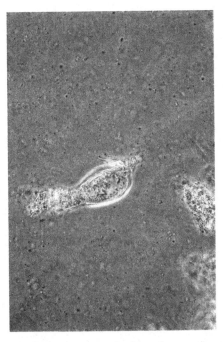

Figure 21.14 *Dispersed rotifer. In the presence of cell bursting agents or harsh surfactants the "soft" cells of the rotifer* Brachinus *are dispersed, while the "hard" or calcified cells of the lorica, foot, and toes are not damaged. All that remains of the rotifer under a microscopic examination is the glowing edge of the lorica as it refracts light from the lamp of the microscope.*

in the operational condition of the treatment process. Active rotifers and free-living nematodes are present in "stable" treatment processes regardless of whether the process has a young sludge age or old sludge age. Inactivity of these organisms commonly occurs due to low dissolved oxygen level, and dispersion of rotifers does occur in the presence of excess surfactants or cell-bursting agents (Figure 21.14).

Bibliography

Al-Ghusain, I. A., and O. J. Hao. 1995. Use of pH as control parameter for aerobic/anoxic sludge digestion. *J. Environ. Eng.* **121**.

Al-Rekabi, W. S., H. Qiang, and W. W. Qiang. 2007. Review on sequencing batch reactors. *Pak. J. Nutrition.* **6**.

Basu, S., S. K. Pilgram, D. W. Keck, and C. Painter. 2006. ORP and pH based control of SBR cycles for nutrient removal from wastewater. *Water Environ. Res.* **78**.

Bungay, S., M. Humphries, and T. Stephenson. 2007. Operating strategies for variable flow sequencing batch reactors. *Water Environ. J.* **21**.

Chambers, B. 1993. Batch operated activated sludge plant for production of high effluent quality at small works. *Water Sci. Tech.* **28**.

Charpentier, J. G. Martin, H. Wacheux, and P. Gilles. 1998. ORP regulation and activated sludge: 15 years of experience. *Water Sci. Tech.* **38**.

Curds, C. R. 1992. *Protozoa in the Water Industry*, Cambridge University Press, Cambridge, U.K.

Curds, C. R. 1969. *An Illustrated Key to the British Freshwater Ciliated Protozoa Commonly Found in Activated Sludge.* Technical Paper No. 12. Water Pollution Resources, Ministry of Technology. London, U.K.

Diamadopoulos, E. P., X. Samaras, and G. Sakellaropulos. 1997. Combined treatment of landfill leachate and domestic sewage in a sequencing batch reactor. *Water Sci. Tech.* **36**.

Gerardi, M. H. 2008. *Microscopic Examination of the Activated Sludge Process*, John Wiley & Sons, Hoboken, NJ.

Gerardi, M. H. 2006. *Wastewater Bacteria*, Wiley-Interscience, Hoboken, NJ.

Gerardi, M. H. 2002. *Nitrification and Denitrification in the Activated Sludge Process*, Wiley-Interscience, New York.

Hamamoto, Y., Tabata, and Y. Okubo. 1997. Development of the intermittent cyclic process for simultaneous nitrogen and phosphorus removal. *Water Sci. Tech.* **35**.

Hoepker, E. C., and E. D. Schroeder. 1979. The effects of loading rate on batch activated sludge effluent quality. *J. Water Polln. Fed.* **51.**

Irvine, R. L., and A. W. Busch. 1979. Sequencing biological reactor—An overview. *J. Water Poll. Cont. Fed.* **51.**

Kargi, F., and A. Uygur. 2003. Nutrient loading rate effects on nutrient removal in a five-step sequencing batch reactor. *Proc. Biochem.* **39.**

Katsogiannis, A. N., M. Kornaros, and G. Lyberatos. 2002. Long-term effect of total cycle time and aerobic/anoxic phase ratio of nitrogen removal in a sequencing batch reactor. *Water Environ. Res.* **74.**

Kim, H., and O. J. Hao. 2001. pH and oxidation-reduction potential control strategy for optimization of nitrogen removal in an alternating aerobic-anoxic system. *Water Environ. Res.* **73.**

Koch, F. A., and W. K. Olham. 1985. Oxidation–reduction potential—A tool for monitoring, control and optimization of biological nutrient removal systems. *Water Environ. Res.* **69.**

Mahvi, A. H., A. R. Mesdaghinia, and F. Karakani. 2004. Feasibility of continuous flow sequencing batch reactor in domestic wastewater treatment. *Am. J. App. Sci.* **1.**

Nakajima, J., and M. Kaneko. 1991. Practical performance of nitrogen removal in small scale sewage treatment plants operated in intermittent aeration mode. *Water Sci. Tech.* **24.**

New England Interstate Water Pollution Control Commission. 2005. *Sequencing Batch Reactor Design and Operational Considerations*, Lowell, MA.

Norcross, K. L. 1992. Sequencing batch reactors—An overview. *Water Sci. Tech.* **26.**

Pavoni, J. L., M. W. Tenney, and W. T. Echelberger, Jr. 1972. Bacterial exocellular polymers and biological flocculation. *J. Water Pollut. Control Fed.* **44.**

Ros, M., and J. Vrtovsek. 2004. The study of nutrient balance in sequencing batch reactor wastewater treatment. *Acta Chim. Solv.* **51.**

Silverstein, J., and E. D. Schroeder. 1983. Performance of SBR activated sludge processes with nitrification/denitrification. *J. Water Pollut. Control Fed.* **55.**

Steinmetz, H. J., and T. G. Schmitt. 2002. Efficiency of SBR technology in municipal wastewater treatment plants. *Water Sci. Tech.* **46.**

Thornberg, D. E., M. K. Nielsen, and K. L. Andersen. 1993. Nutrient removal: On-line measurements and control strategies. *Water Sci. Tech.* **28.**

U.S. EPA. 1992. Sequencing Batch Reactors for Nitrification and Nutrient Removal, U.S. Environmental Protection Agency, EPA 832 R-92-002, Washington, D.C.

Wareham, D. G., D. S. Marvinic, and K. J. Hall. 1994. Sludge digestion using ORP-regulated aerobic anoxic cycles. *Water Res.* **28.**

Wareham, D. G., K. J. Hall, and D. S. Mavinic. 1993. Real-time control of aerobic–anoxic sludge digestion using ORP. *J. Environ. Eng.* **119.**

Williams, T. M., and R. F. Unz. 1983. Environmental distribution of Zoogloea strains. *Water Res.* **17.**

Yu, R. F., S. L. Chang, W. Y. Cheng. 1998. Applying real-time control to enhance the performance of nitrogen removal in the continuous-flow SBR system. *Water Sci. Tech.* **38.**

Glossary

Acclimate   Repair or damage enzymes or production of new enzymes.

Aerobic   Environment having free molecular oxygen.

Agglutination   Joining together of bacterial cells to form a floc particle or biofilm.

Aliphatic   Open-chain compounds.

Ammonification   Release of amino groups from the biological degradation of organic-nitrogen compounds.

Amorphous   Without specific form or shape.

Anaerobic   Environment that does not have free molecular oxygen.

Anion   A negatively charged ion.

Anoxic   Environment having only nitrate.

Aromatic   Benzene derivatives.

Assimilation   Incorporation of nutrients into cellular material.

Bioaugmentation   The addition of microbial cultures (bacteria and fungi) to wastewater treatment facilities.

Biochemical   Chemical reactions occurring inside living cells.

Biomass   The weight of all organisms in an environment.

Carbonaceous   Organic compounds or compounds that contain carbon and hydrogen.

Cation   A positively charged ion.

Cellulose   The most complicated, insoluble sugar complex that forms the cell walls in all plants.

Troubleshooting the Sequencing Batch Reactor, by Michael H. Gerardi
Copyright © 2010 by John Wiley & Sons, Inc.

Cilia Short hair-like structures that are found on the surface of protozoa and metazoa that beat in unison to provide locomotion and collect dispersed bacterial cells as substrate.

Denature Damage to enzymatic (protein) structure resulting in loss of enzymatic activity.

Dendritic Bearing markings that are treelike.

Diurnal Cyclic changes during the day.

Electrolyte Chemical or its solution in water, which conducts current through ionization.

Endogenous Degradation of stored food.

Endoenzyme Enzyme produced inside the cell that remains inside the cell and degrades soluble substrate.

Enzyme Proteinaceous molecule that accelerates the rate of a biochemical reaction without being consumed during the reaction.

Eutrophication Rapid aging of a body of water caused by accelerated growth of aquatic plants from nutrient-rich influent.

Exoenzyme An enzyme produced inside the cell that is released through the cell membrane and cell wall and solubilizes substrate that is adsorbed to the cell.

Facultative An organism that is anaerobic but has the ability to use free molecular oxygen, if oxygen is available.

Fermentative Anaerobic condition resulting in the production of acids and alcohols from the degradation of organic compounds and the production of hydrogen sulfide and/or sulfide from the reduction of sulfate.

Fibril Short (<2–5 nm) projections from the cell membrane through the cell wall of bacteria that are used in cellular agglutination.

Field of view The circular area of view when looking through the microscope.

Filtrate The liquid portion of a sample that passes through filter paper.

Flocculant Chemical compound added to a waste stream to capture and thicken solids.

Gelatinous Apparently solid, often jellylike, material formed from a colloidal solution.

Hydrolysis The decomposition of compounds by interaction with water.

Hydrophobic Not soluble in water. Separates from water.

In situ In place or on-site.

Inert Nonliving.

Intracellular Within the cell.

Ion Any atom or molecule that has an electric charge due to loss or gain of electrons.

Kinetic Energy obtained from biochemical reactions.

Lipid Generic term for oils, fats, waxes, and related products found in living tissues.

Lysis Decomposition or splitting of cells or molecules.

Metazoa A subkingdom of the animal kingdom, comprising multicellular animals having two or more tissues layers and showing a high degree of coordination between the different cells compromising the body.

Methemoglobinema   "Blue baby disease." Disease caused by the consumption of nitrate-laden groundwater.

Nitrogenous   Nitrogen-containing compounds.

Organotroph   An organism that obtains it carbon and energy from organic compounds.

Oxic   Environment having free molecular oxygen.

Polyhydroxybuytrate   One of many insoluble starch granules produced by bacterial cells.

Polysaccharide   A group of complex carbohydrates such as starch and cellulose.

Protein   A large organic molecule containing amino acids that have amino groups (-NH$_2$) and carboxyl groups (-COOH).

Selector   A tank, zone, or period of time that establishes an operational condition that favors (selects for) the growth of floc-forming bacteria and disfavors (selects against) the growth of filamentous organisms.

Substrate   Carbon and energy sources (food) for organisms.

Xenobiotic   Foreign to the body or manmade compounds.

Abbreviations and Acronyms

AOB	Ammonia-oxidizing bacteria
BNR	Biological nutrient removal
BOD	Biochemical oxygen demand
BOD_u	Ultimate, biochemical oxygen demand
BOD_5	5-day, biochemical oxygen demand
BOD_7	7-day, biochemical oxygen demand
BOD_{21}	21-day, biochemical oxygen demand
cBOD	Carbonaceous, biochemical oxygen demand
COD	Chemical oxygen demand
CSO	Combined sewer overflow
°C	Degree Celsius
EBPR	Enhanced biological phosphorus removal
F/M	Food-to-microorganism ratio
FOG	Fats, oils, and grease
g	gram
hr	hour
HTH	High-test hypochlorite
I/I	Inflow and infiltration
MCRT	Mean cell residence time
MGD	Million gallons per day
MLSS	Mixed-liquor suspended solids
MLVSS	Mixed-liquor volatile suspended solids

Troubleshooting the Sequencing Batch Reactor, by Michael H. Gerardi
Copyright © 2010 by John Wiley & Sons, Inc.

mg/L	Milligram per liter
MSDS	Material safety data sheet
mV	millivolt
nBOD	Nitrogenous biochemical oxygen demand
NOB	Nitrite-oxidizing bacteria
O & M	Operation and maintenance
ORP	Oxidation–reduction potential
OUR	Oxygen uptake rate
PAC	Poly-aluminum chloride
PAO	Phosphorus-accumulating organism
Redox	Oxidation–reduction potential
RR	Respiration rate
SBR	Sequencing batch reactor
SOUR	Specific oxygen uptake rate
spp.	species
SRB	Sulfate-reducing bacteria
SVI	Sludge volume index
TKN	Total Kjeldahl nitrogen
TDS	Total dissolved solids
TSS	Total suspended solids
WHO	World Health Organization
μmohm/cm	Micro-ohms per centimeter

Index

Acid-forming bacteria 22. 105, 123, 155, 174, 178
Aerated fill 6, 19, 21
Aerobic 3, 8, 11, 27, 62, 63
Alkalinity 9, 10. 49, 53, 56, 58, 59, 65, 74, 96, 99–111, 156, 158, 163, 164, 165, 177, 178, 180
Ammonification 35, 66, 68, 144, 146, 147–148
Anaerobic/fermentative 3, 8, 11, 16, 21, 22, 23, 27, 65, 70, 94, 100, 101, 104, 105, 110, 111, 129, 147, 148, 150, 155, 156, 162, 163, 171, 178, 180

Benzene 41, 43
Bioaugmentation 10, 14, 15, 78, 92, 95, 96, 97, 120, 121
Biological phosphorus release 8, 10, 21, 22, 64, 65, 94, 104, 119, 124, 150, 151, 155–158, 162, 177, 178
Biological phosphorus uptake 21, 22, 23, 64, 123, 150, 164, 177, 178

Biochemical oxygen demand (BOD) 7, 9, 10, 13, 14, 15, 16, 21, 22, 23, 24, 29, 30, 33–40, 41, 42, 43, 44, 49, 52, 53, 54, 55, 56, 57, 58, 59, 61, 62, 62, 64, 65, 66, 67–78, 81, 82, 83, 84, 86, 90, 92, 97, 103, 104, 109, 110, 113, 115, 117, 121, 123, 126, 129, 130, 131, 135, 136, 138, 140, 141, 155, 156, 164, 167, 168, 169, 171, 172, 173, 174, 175, 176, 177, 180, 186, 188

Chlorine sponge 54
Chemical oxygen demand (COD) 33, 41–45
Conductivity 73, 111
Conventional, activated sludge process 3, 4, 7, 8, 9, 10, 17, 18
Cold weather 13, 17, 54, 55, 58, 65

Denitrification 8, 10, 13, 15, 21, 23, 24, 25, 49, 52, 61–66, 72, 83, 101, 102, 103, 106, 109, 128, 144, 147–148, 154, 163, 165, 177, 178, 179, 180

Troubleshooting the Sequencing Batch Reactor, by Michael H. Gerardi
Copyright © 2010 by John Wiley & Sons, Inc.

Printed in the United States
By Bookmasters